AF266030

Published by:

Pietas Publications
Waynesboro, Virginia, USA
web: www.jasperburns.com
email: pietas@jasperburns.com

VIRGINIA THROUGH TIME

A Natural History Atlas

Black and White Edition

By Jasper Burns

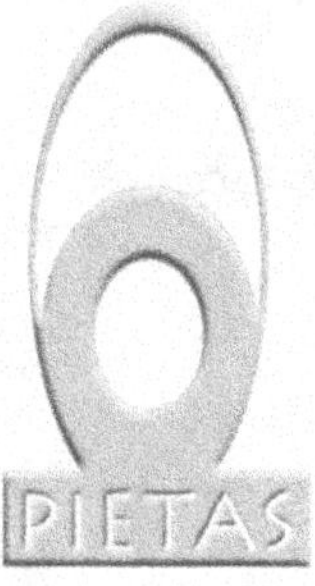

Copyright 2014

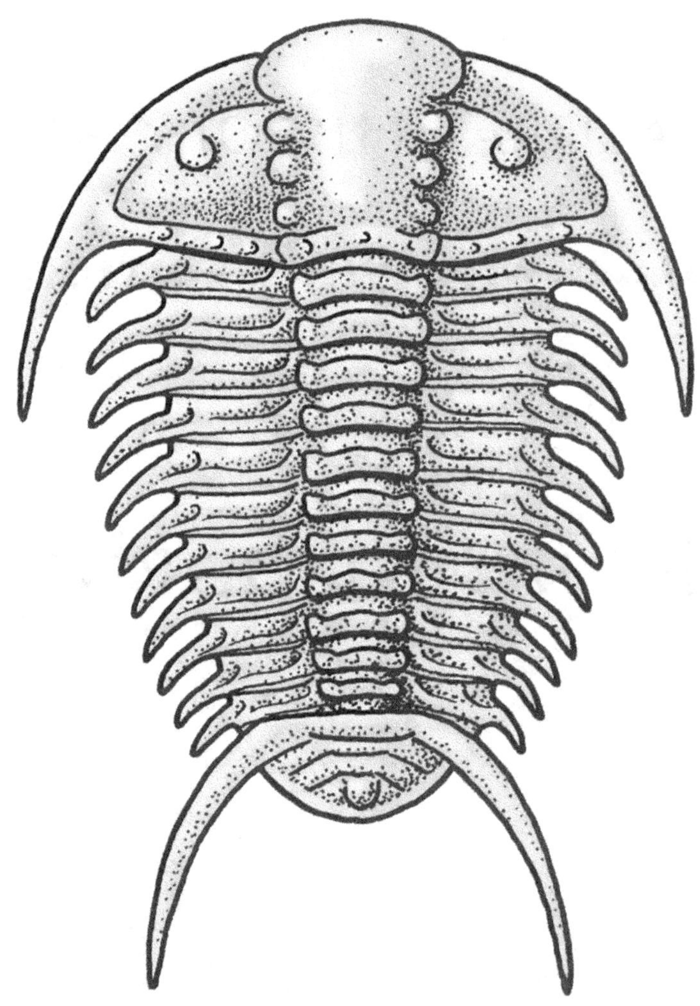

TRILOBITE - *Ceraurus (Ordovician Period)*

Virginia Through Time

Acknowledgements

Most of the illustrations in this book are based on drawings originally prepared by the author for the following Virginia Museum of Natural History publications: "Exploring Fossils", by Jasper Burns, "Fun With Mammals", by Nancy Moncrief, "Model Inquiries into Nature in the Schoolyard: The MINTS Book - An Inquiry Field Guide to the Natural History of Southwestern Virginia Schoolyards", by Frank Taylor, Alan Raflo, and Llyn Sharp, and "Exploring Virginia's Endangered Species", by Tracy Triggs. They are used by permission. The drawings of the gastropods *Ecphora gardnerae* and *Ecphora quadricostata* are used with the permission of Dr. Eric J. Seifter, MD, for whom they were prepared.

Thanks are due to Dr. Lauck W. Ward and Dr. Nicholas Fraser of the Virginia Museum of Natural History for their advice on the reconstructions of the prehistoric life of the Tertiary and Triassic-Jurassic-Cretaceous periods respectively. Assistance was also provided by Alan Raflo, Llyn Sharp, and Nancy Moncrief during the preparation of the drawings of present-day plants and animals. Rick Boland deserves credit for supervising and encouraging the publication of the VMNH books, all of which are now out of print.

Introduction

The remnants of Virginia's past may be found in the layers of rock and soil that support the plants and animals that live here today. From bottom to top, these layers can be read like the pages of a book. Its illustrations are the fossils, rocks, artifacts, and geologic structures. Anyone willing to go outside and take a look can read the story. (The author's book, "Fossil Collecting in the Mid-Atlantic States" (Johns Hopkins University Press, 1991) describes numerous places where fossils can be found in Virginia and neighboring states.)

Virginia is fortunate in having a rich fossil record for at least one of the ancient environments that existed in almost every major division (eras, periods, and epochs) of geologic time since fossils became commonly preserved in the Cambrian Period, more than 500 million years ago. The only exceptions are the Permian Period (299-252 million years ago) and the Oligocene Epoch (33.9-23.0 million years ago). On the following pages are reconstructions of some of these environments with maps showing (1) what the region looked like at the time in terms of elevation and coastlines and (2) where sedimentary deposits of that age are exposed today. In addition, drawings are included of sample fossils and of representative plants and animals of the present time. (**Note:** the illustrations are not drawn to the same scale so that larger specimens can share pages with smaller ones.)

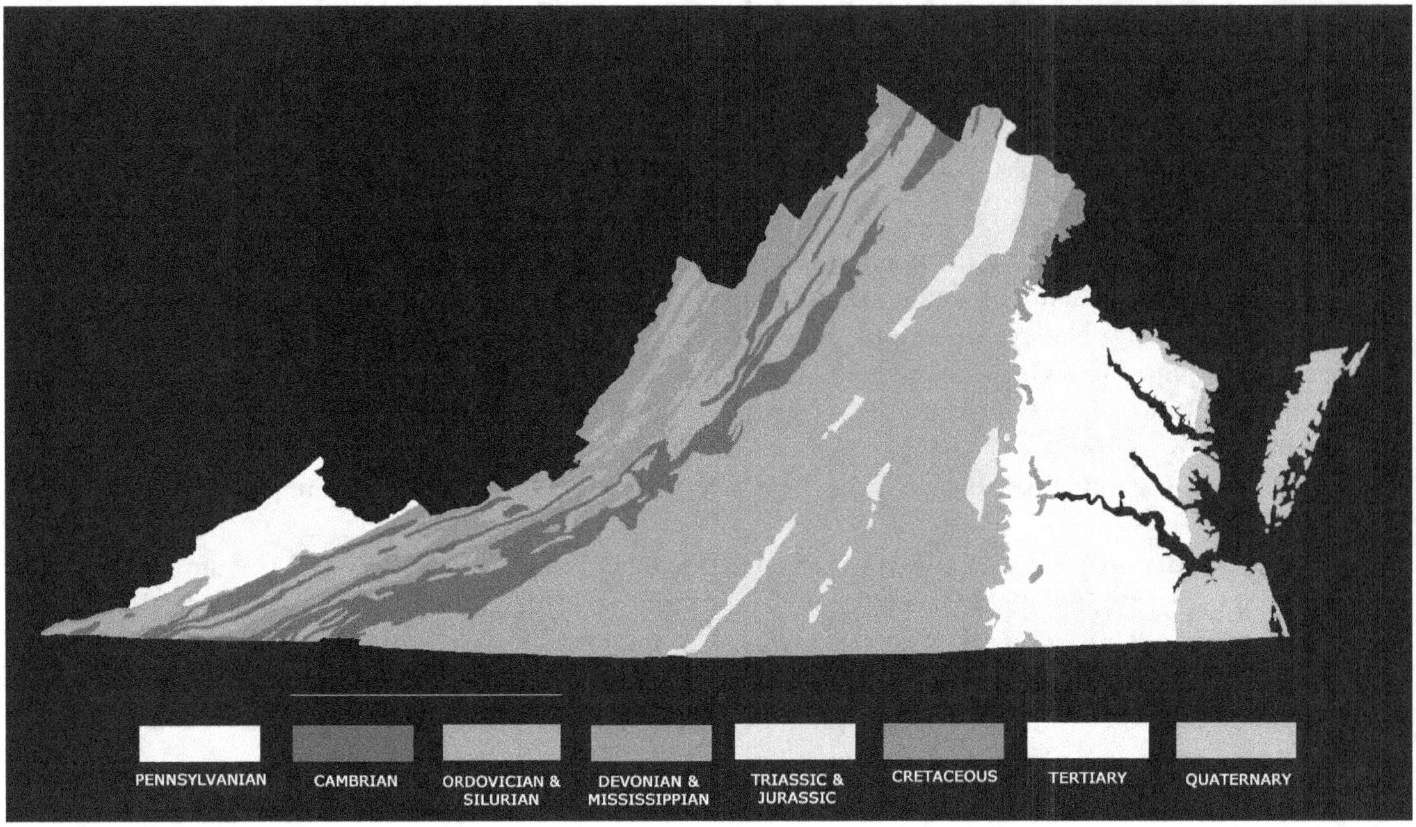

Simplified geologic map of Virginia showing the age of surface sedimentary rocks and deposits. The gray areas signify igneous and metamorphic rocks that contain few if any recognizable fossils.

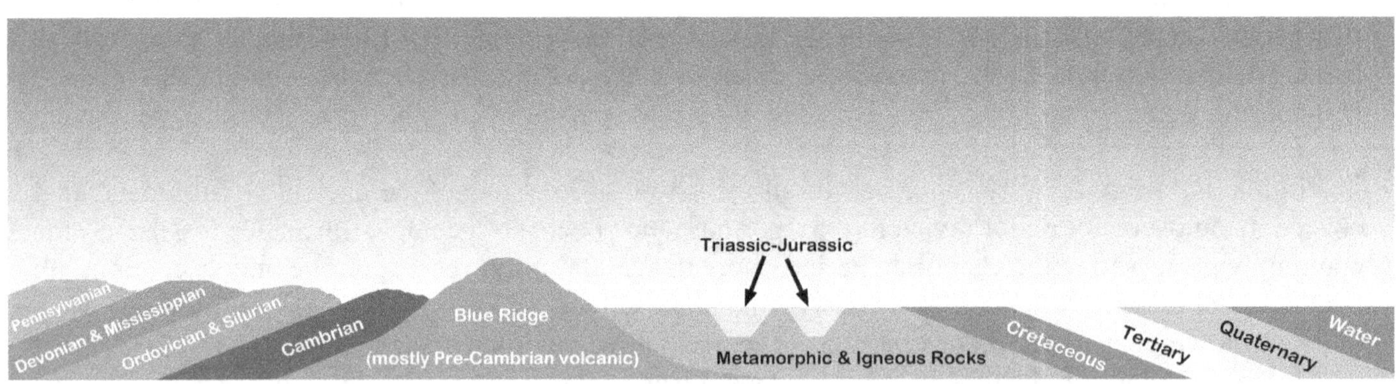

Idealized cross-section of the rocks of Virginia. Notice that younger rocks lay on top of older rocks. The situation in the western part of the state is more complicated than shown because of the folding of rock layers during mountain building and differential rates of erosion. However, the sequence of layers is clear, with rocks generally tending to be younger the further west one goes from the Blue Ridge. The opposite is true in eastern Virginia, with sediments growing younger as one moves *east*, reflecting the gradual retreat eastward of the coastline during most of the Cenozoic Era.

GEOLOGIC TIME SCALE

(TIME OF ABUNDANT FOSSILS)

CENOZOIC ERA - 66 mya* to present

    Quaternary Period - 2.6 mya to present
        Holocene Epoch - 11,700 years ago - present
        Pleistocene Epoch - 2.6 mya to 11,700 years ago

    Tertiary Period - 66 to 2.6 mya
      Neogene
        Pliocene Epoch - 5.3 to 2.6 mya
        Miocene Epoch - 23.0 to 5.3 mya
      Paleogene
        Oligocene Epoch - 33.9 - 23.0 mya
        Eocene Epoch - 56 to 33.9 mya
        Paleocene Epoch - 66 to 56 mya

MESOZOIC ERA - 252 to 66 mya

    Cretaceous Period - 145 to 66 mya
    Jurassic Period - 201 to 145 mya
    Triassic Period - 252 to 201 mya

PALEOZOIC ERA - 541 to 252 mya

    Permian Period - 299 to 252 mya
    Pennsylvanian Period - 323 to 299 mya
    Mississippian Period - 359 to 323 mya
    Devonian Period - 419 to 359 mya
    Silurian Period - 444 to 419 mya
    Ordovician Period - 485 to 444 mya
    Cambrian Period - 541 to 485 mya

** mya = millions of years ago*

The Cambrian Period in Virginia
(541-485 million years ago)

In the early Cambrian Period, the bit of Earth's crust that is now western Virginia (whose story we will follow in the following pages) was south of the Equator and turned 90 degrees clockwise from its present position. What is now the Blue Ridge was part of an offshore group of volcanic islands, and the future eastern Virginia was many miles further east and attached to what is now Africa. As the Paleozoic Era progressed, the various building blocks of the state moved north and closer together, finally joining each other by the end of the era.

Western Virginia was covered by a shallow tropical sea in the Cambrian that teemed with invertebrate animals, including snails (gastropods) and bivalved brachiopods as well as extinct animals such as trilobites, archaeocyathids, and hyolithids. Marine plants also thrived, including blue-green algae that formed large, reef-like structures known as stromatolites.

The illustration to the left shows a marine community of the early Cambrian based on fossils from the rock formation known as the Shady Dolomite. These are among the oldest fossils in Virginia, though some algae and "worm tube" fossils (*Skolithus*) from the earliest Cambrian are even older.

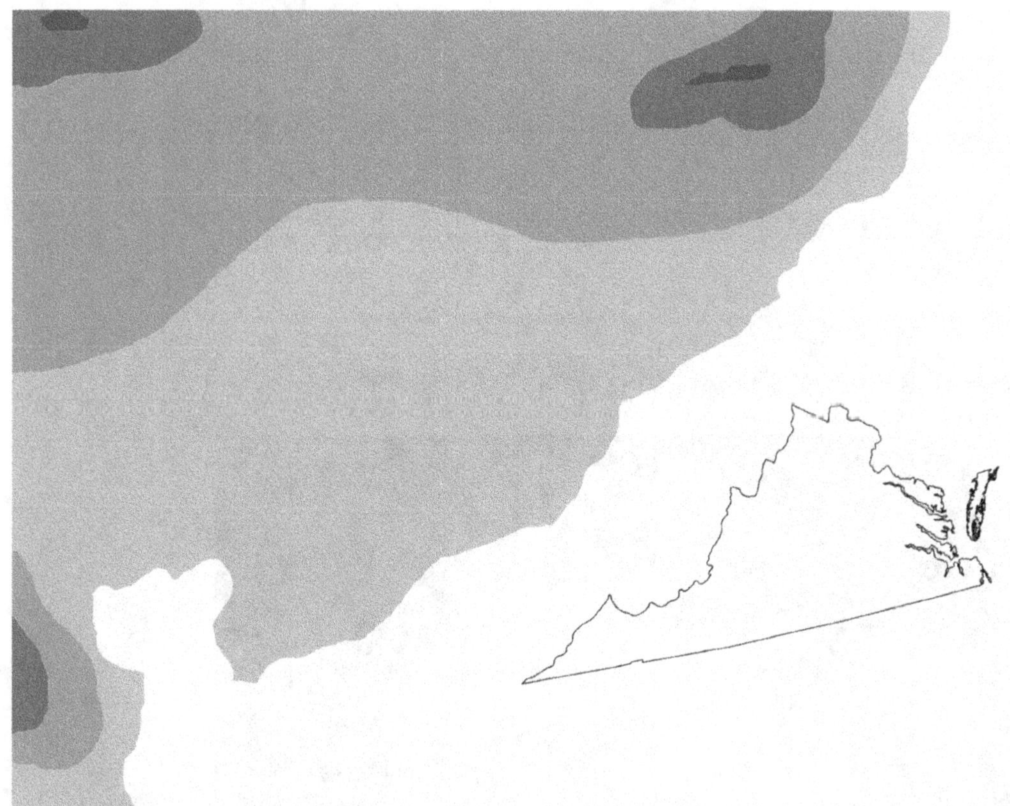

Elevation map of the region of Virginia (outline) during the Cambrian Period, with white representing water and progressively darker shades of gray representing increasing land elevations. (After a map by Dr. Ron Blakey, Northern Arizona University.)

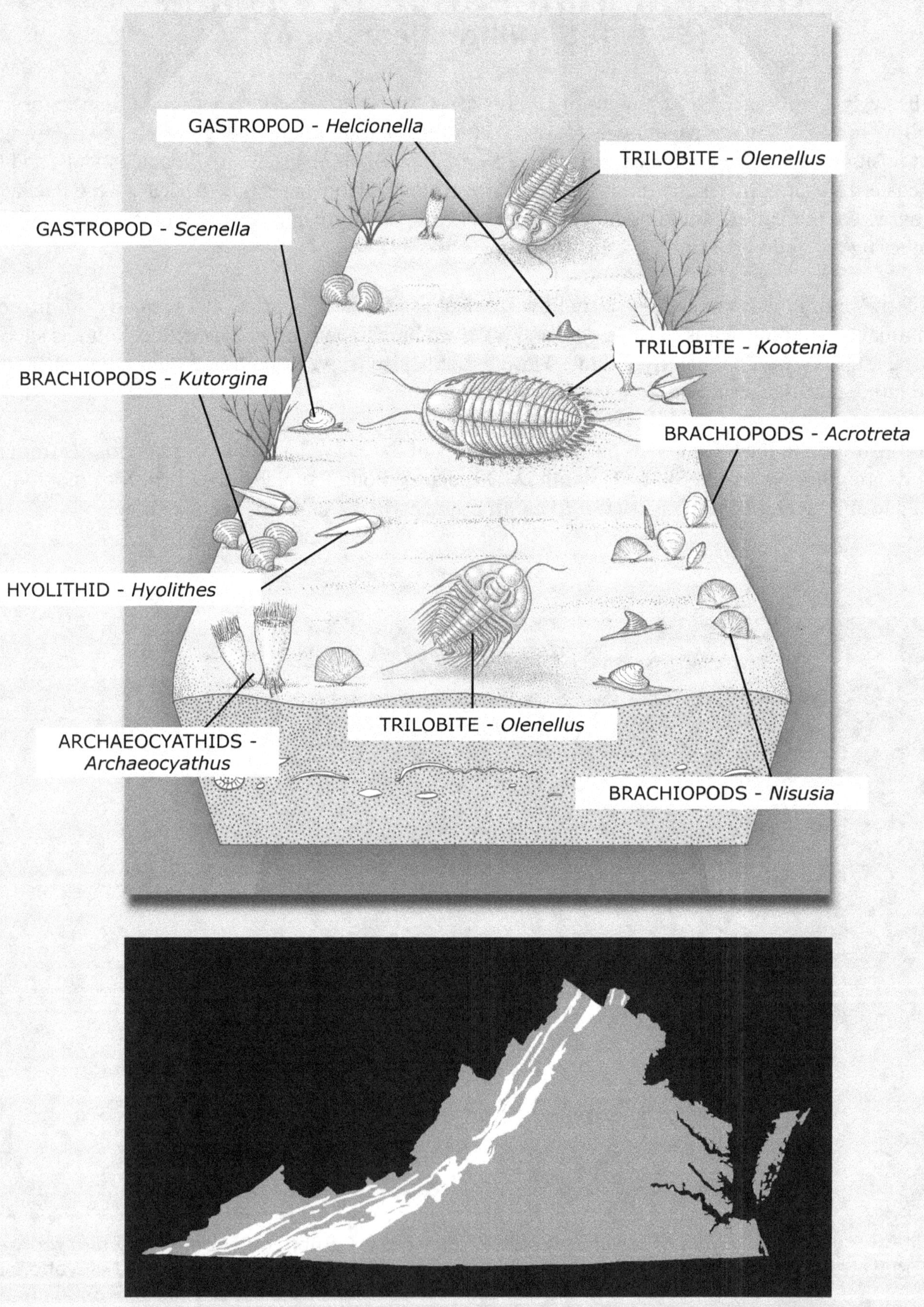

Distribution of surface Cambrian sedimentary rocks in Virginia (white)

Cambrian Fossils of Virginia

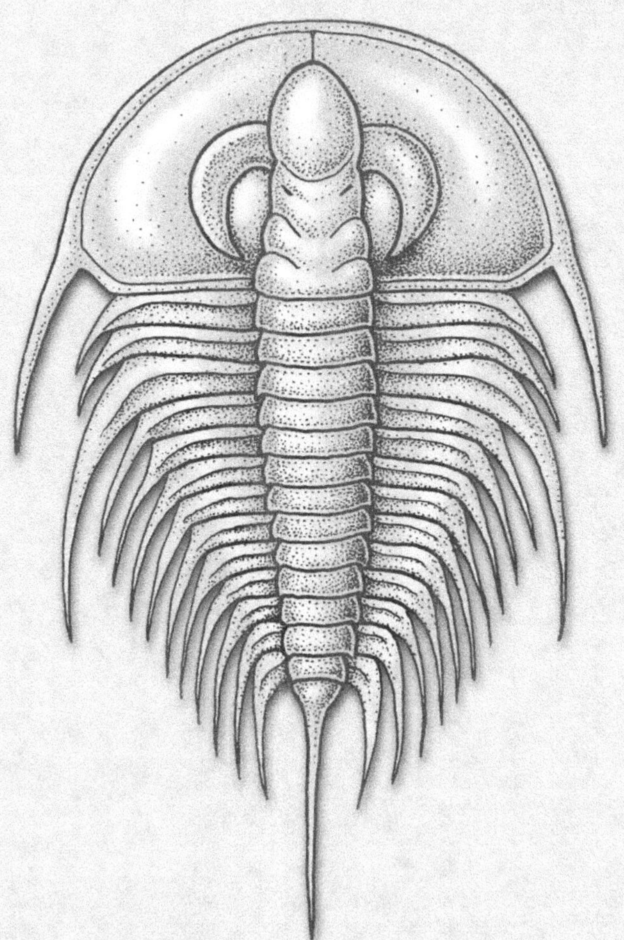

TRILOBITE - *Olenellus*

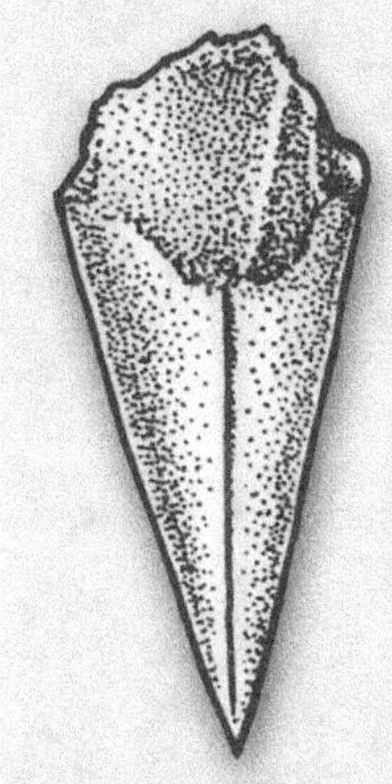

HYOLITHID - *Hyolithes*

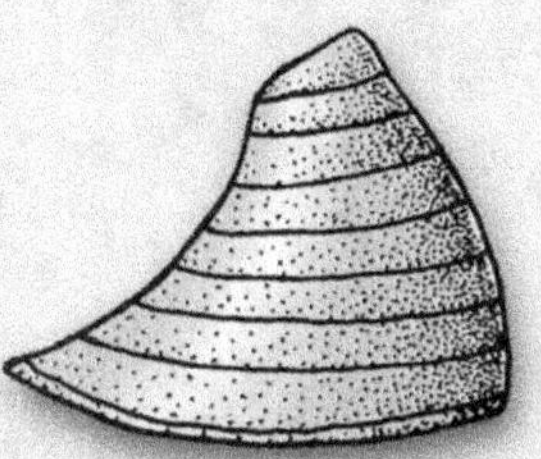

GASTROPOD (snail) - *Helcionella*

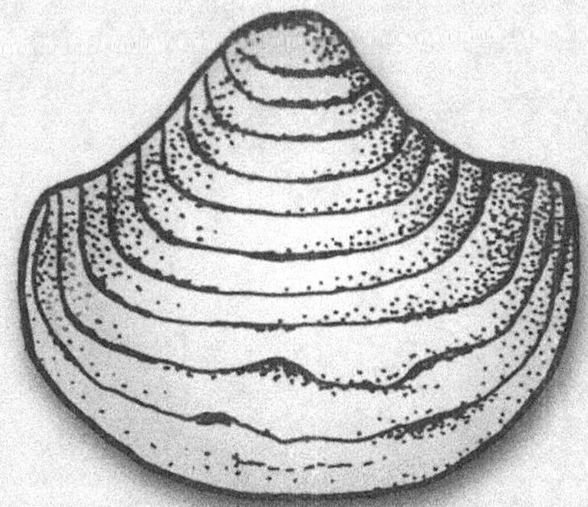

BRACHIOPOD - *Kutorgina*

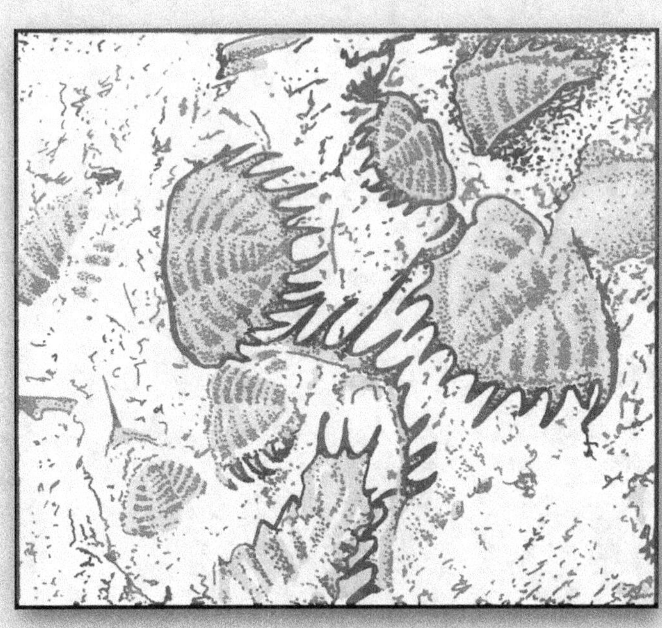

TRILOBITE - *Kootenia* tails (pygidia)

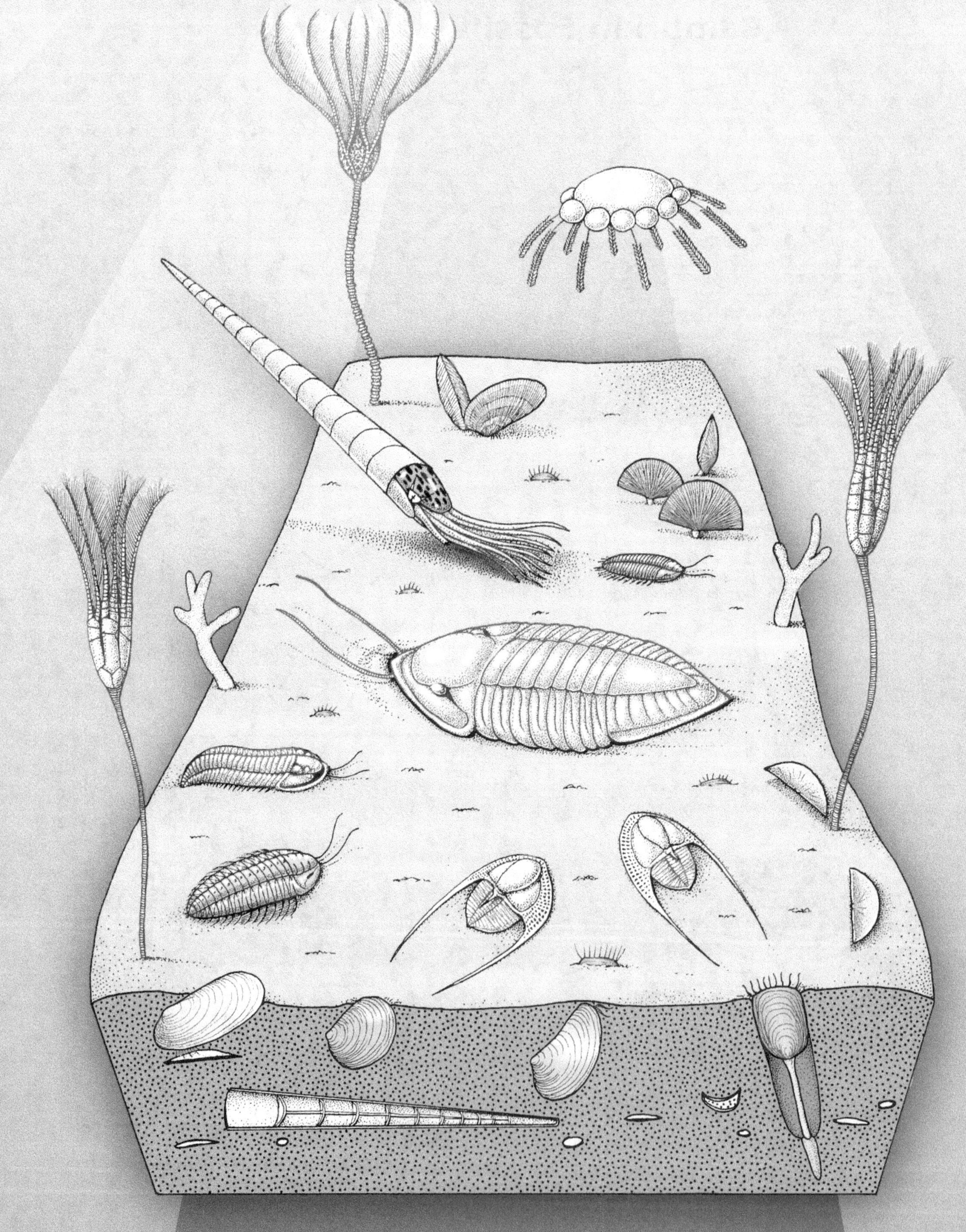

The Ordovician Period in Virginia
(485-444 million years ago)

The early Ordovician in Virginia featured shallow seas in which limestones were deposited in an environment similar to that of the present-day Bahamas. However, as the period progressed, mountain-building began in the area. As some of the land rose above sea level, large quantities of mud and sand were washed out to sea that would eventually form shales and sandstones.

The sea life of the Ordovician featured several new creatures that had not been present during the Cambrian. These included nautiloid mollusks that were related to octupuses and squids and grew long, cone-shaped cells. The "pearly Nautilus" is a living example of this group. Other residents of the Ordovician seas of Virginia were trilobites, brachiopods, crinoids and cystoids (related to starfish), graptolites (extinct), corals, colonial bryozoans, sponges, gastropods (snails), bivalves (clams and their kin), and early fish.

The re-creation at left shows a marine community of the late Ordovician (457-445 mya) based on fossils found in the Martinsburg Formation, which consists primarily of shale.

Elevation map of the region of Virginia (outline) during the Ordovician Period, with white representing water and progressively darker shades of gray representing increasing land elevations. (After a map by Dr. Ron Blakey, Northern Arizona University.)

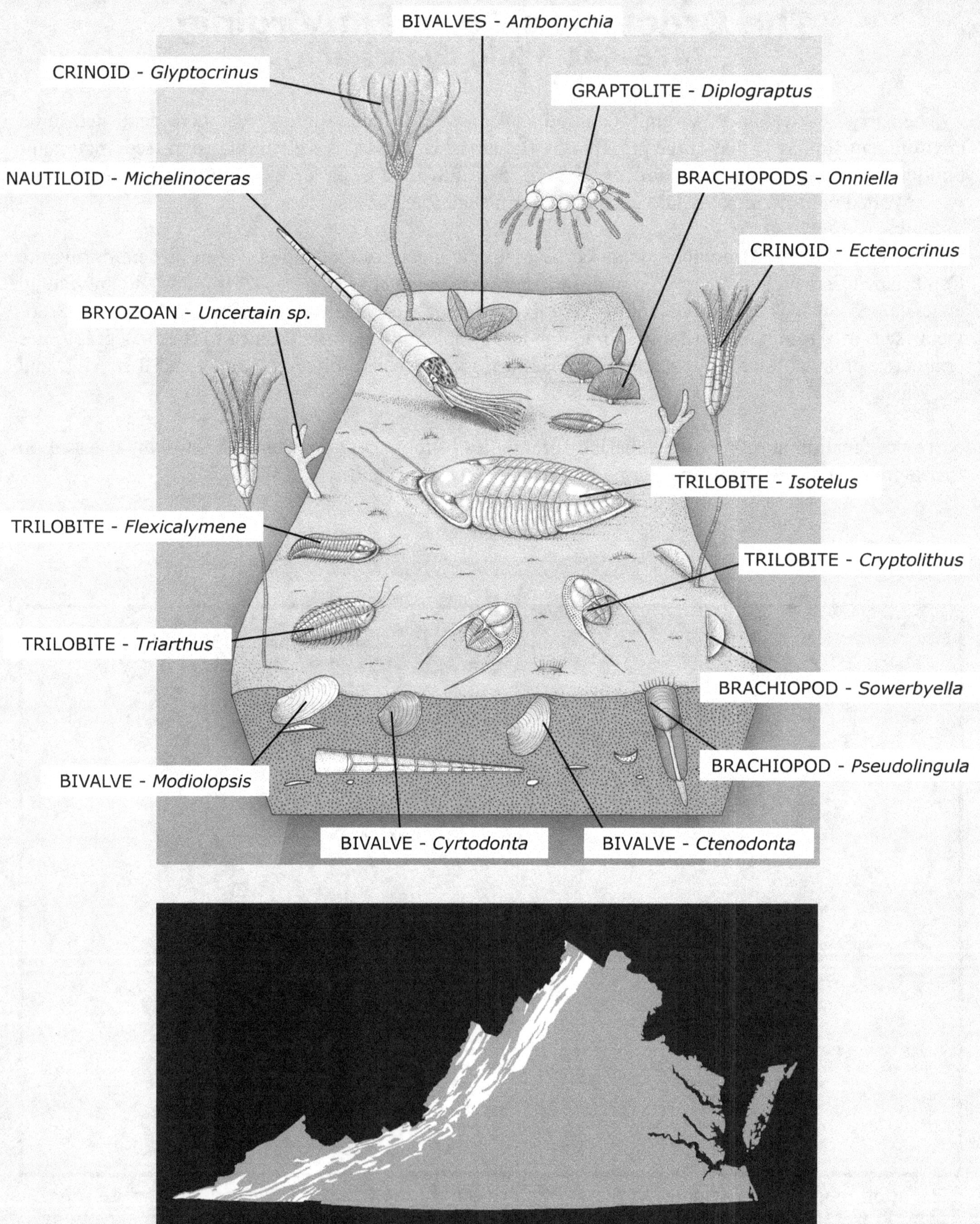

Distribution of surface Ordovician sedimentary rocks in Virginia (white)

Ordovician Fossils of Virginia

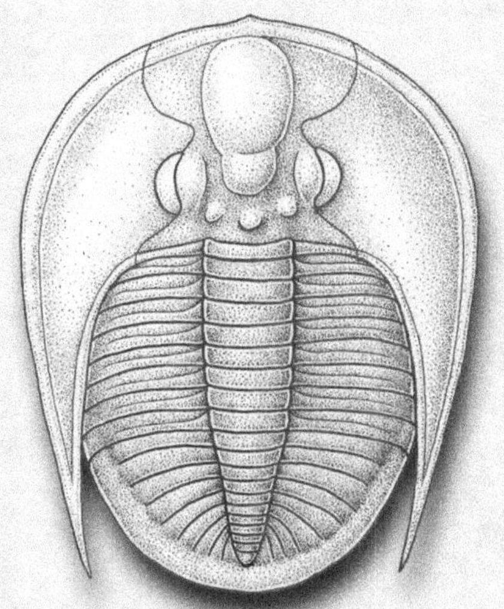

TRILOBITE - *Basiliella*

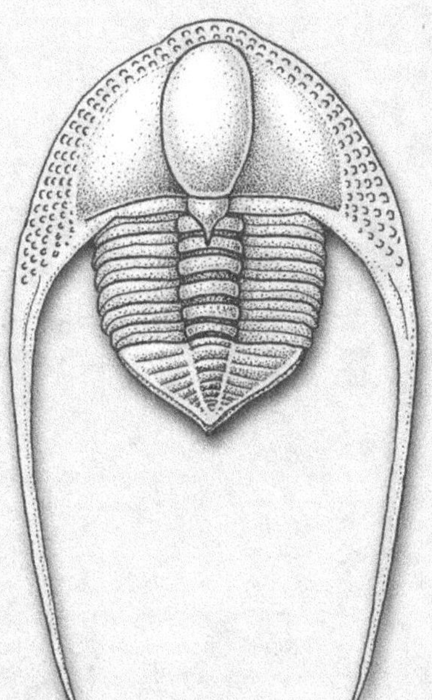

TRILOBITE - *Cryptolithus*

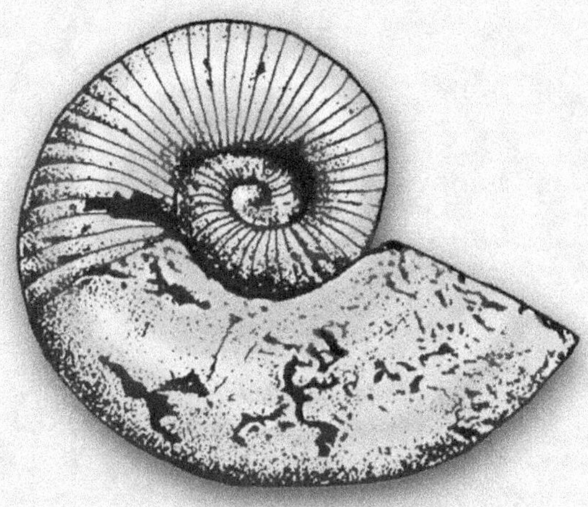

NAUTILOID - *Campbelloceras*

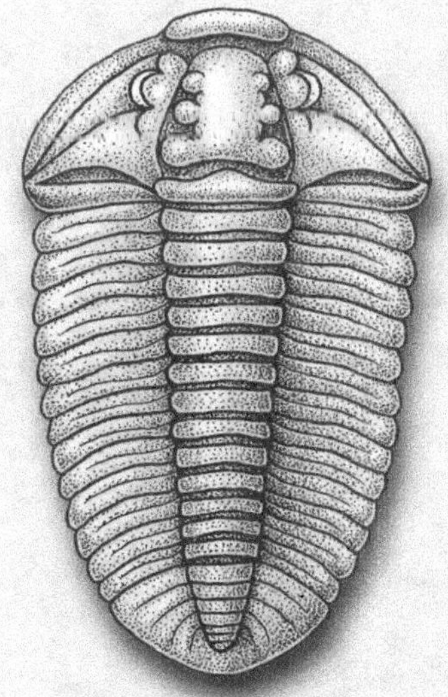

TRILOBITE - *Ampyxina*

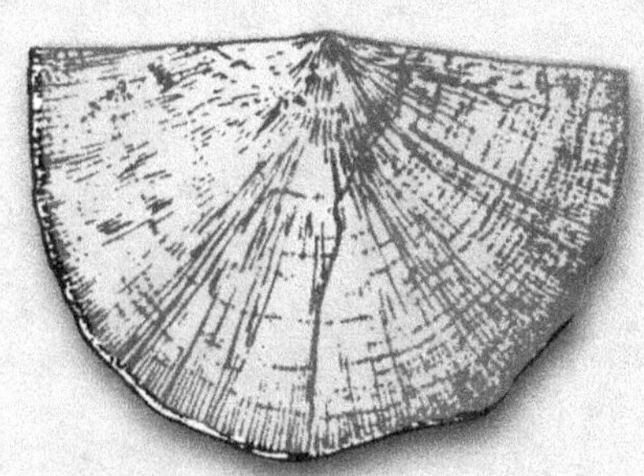

BRACHIOPOD - *Rafinesquina*

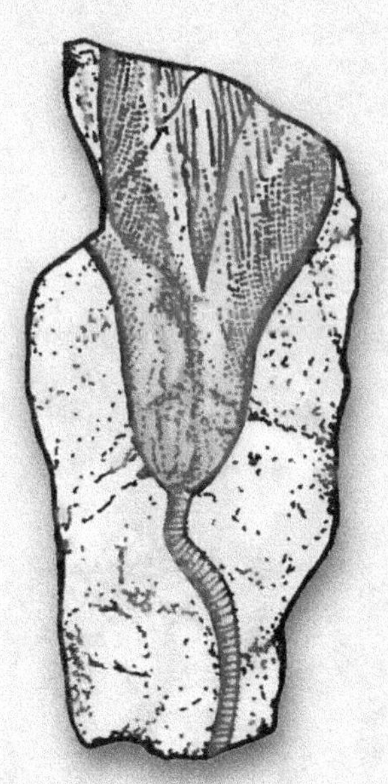

CRINOID

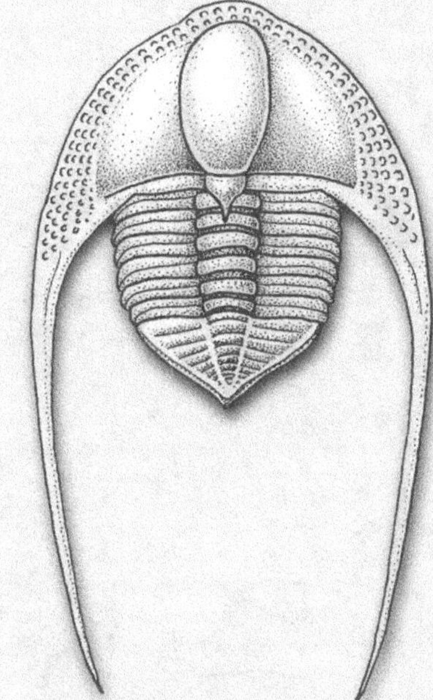

TRILOBITE - *Flexicalymene*

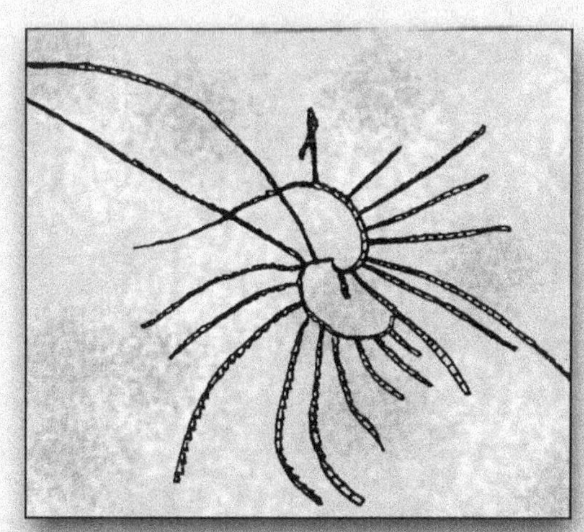

GRAPTOLITE - *Nemagraptus*

The Silurian Period in Virginia
(444-419 million years ago)

The mountain-building at the end of the Ordovician Period continued into the early Silurian and great thicknesses of sandstone were deposited. However, as the period progressed, the mountains were worn down and less sediment was carried out to sea. As a result, limestone formation increased in the later Silurian. These rocks contain an abundance of fossils that are similar to those found in the Ordovician. Trilobites were still common, as were brachiopods, corals, crinoids, bryozoans, gastropods, nautiloids, and bivalves. Crinoid-like cystoids thrived during this period and their fossils are common in Virginia.

The scene of underwater Silurian life opposite is based on fossils found in the Keyser Limestone.

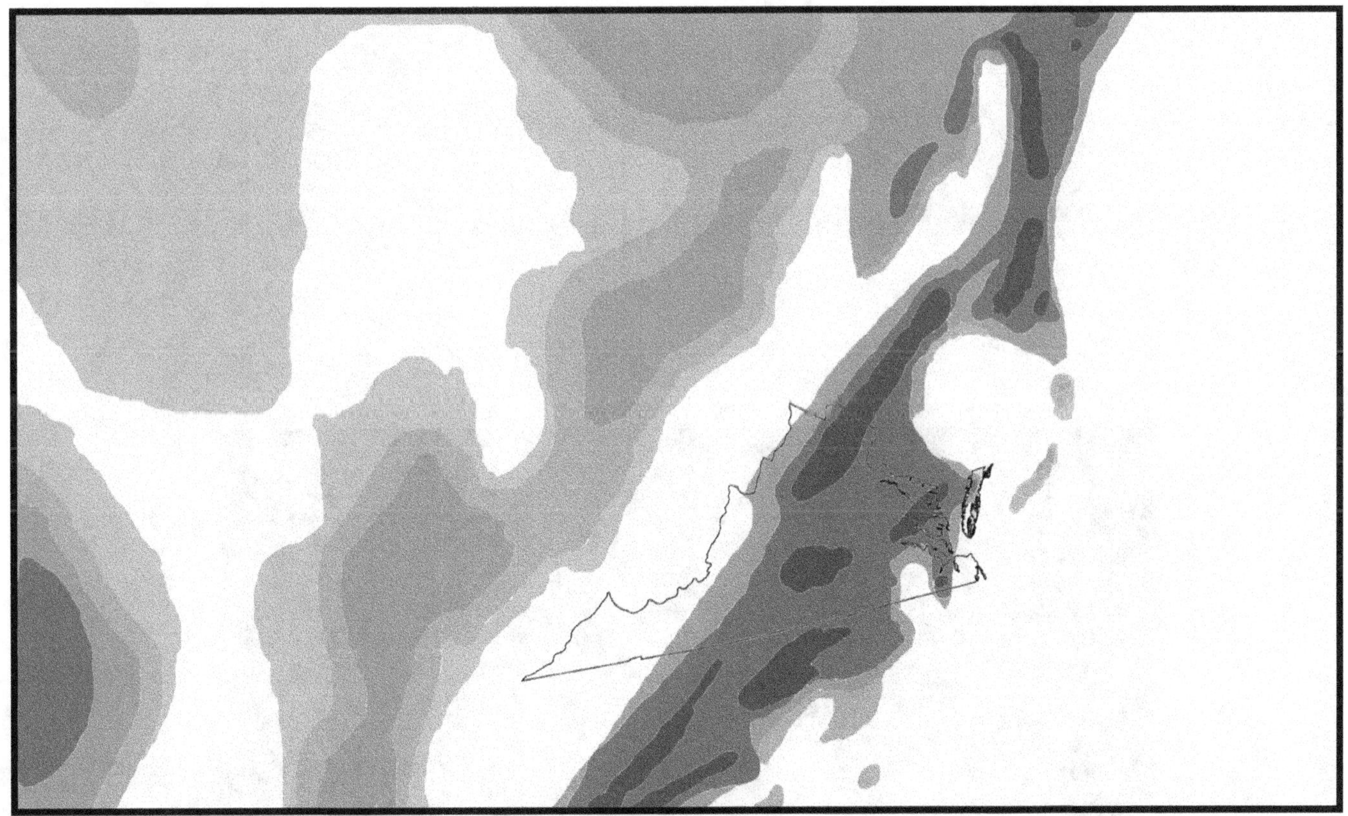

Elevation map of the region of Virginia (outline) during the Silurian Period, with white representing water and progressively darker shades of gray representing increasing land elevations.
(After a map by Dr. Ron Blakey, Northern Arizona University.)

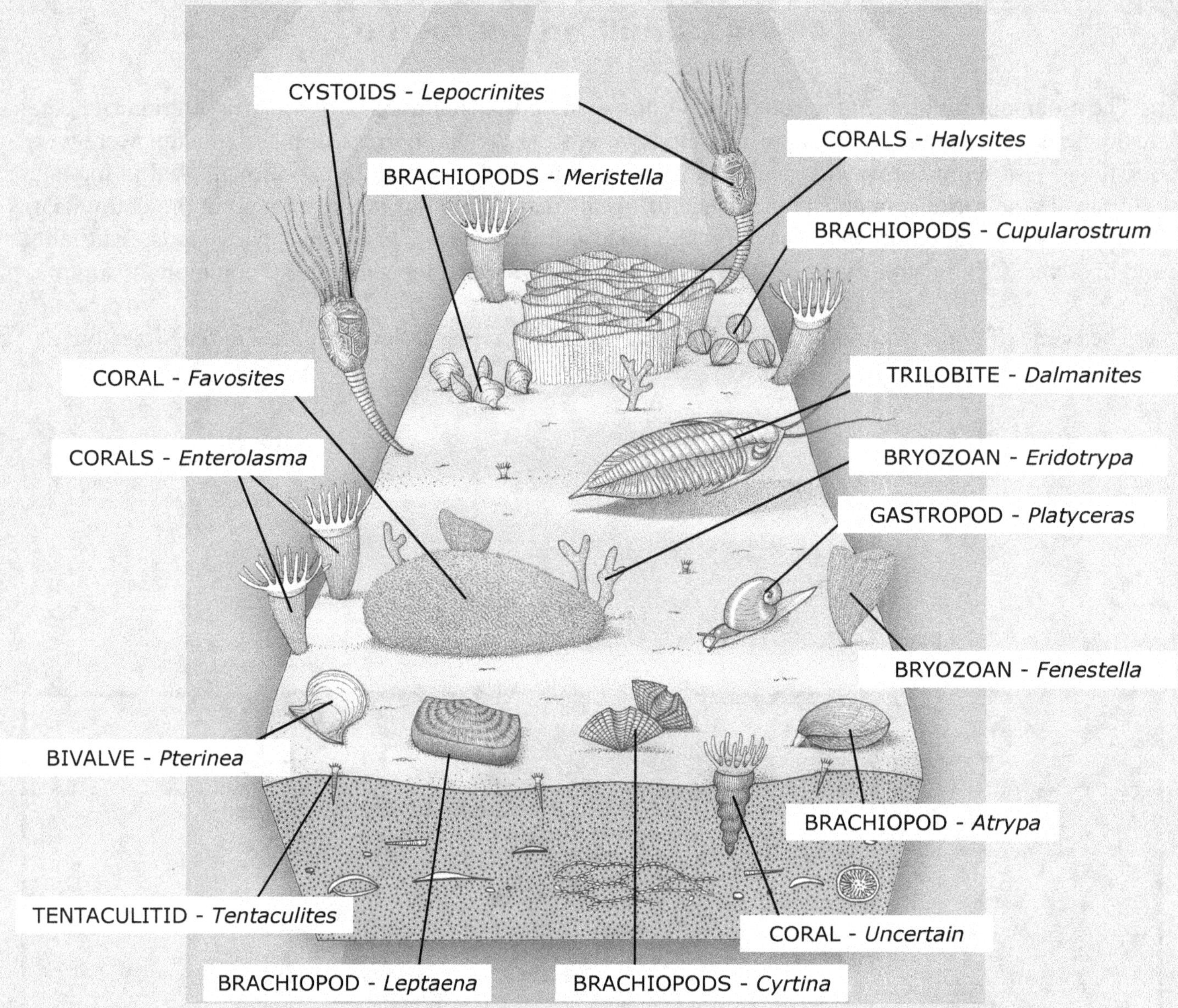

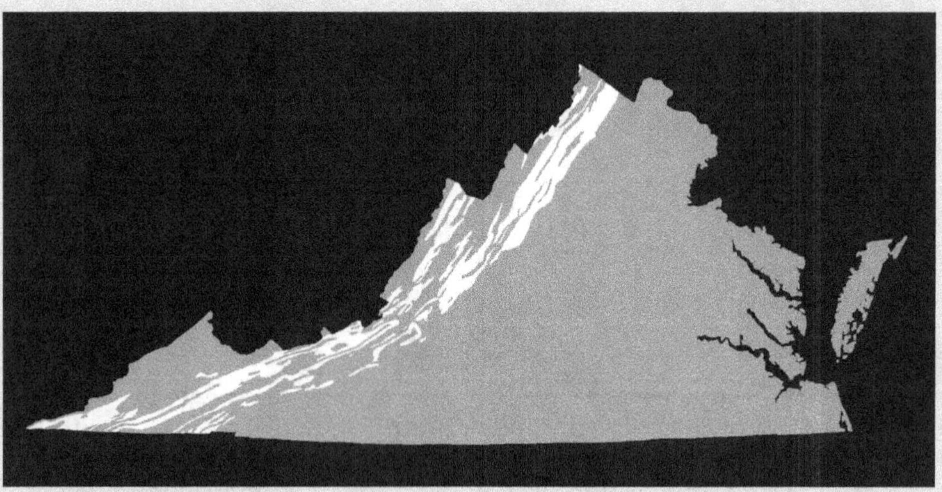

Distribution of surface Silurian sedimentary rocks in Virginia (white)

Silurian Fossils of Virginia

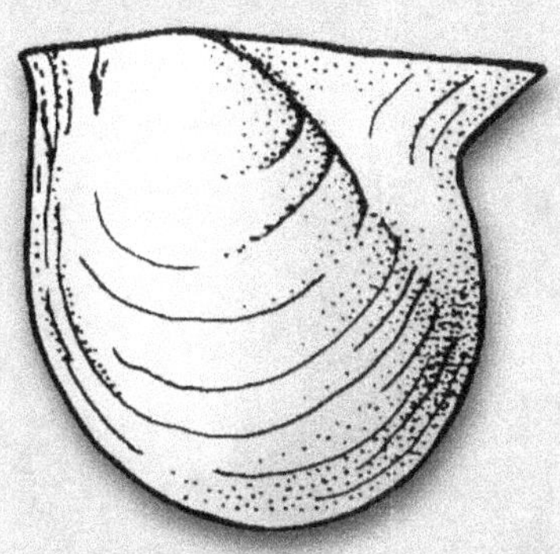

BIVALVE - *Pterinea*

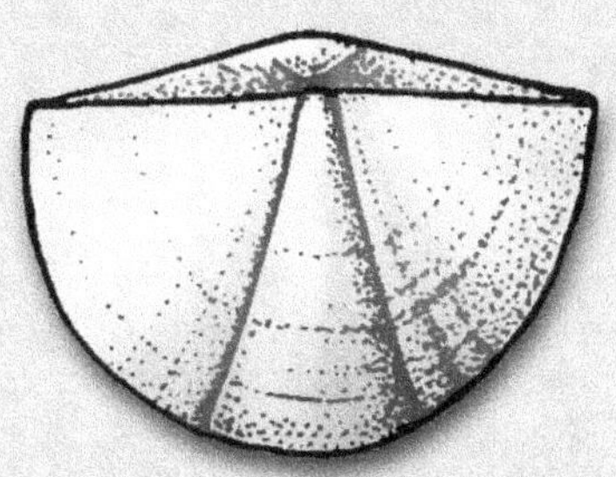

BRACHIOPOD - *Reticularia*

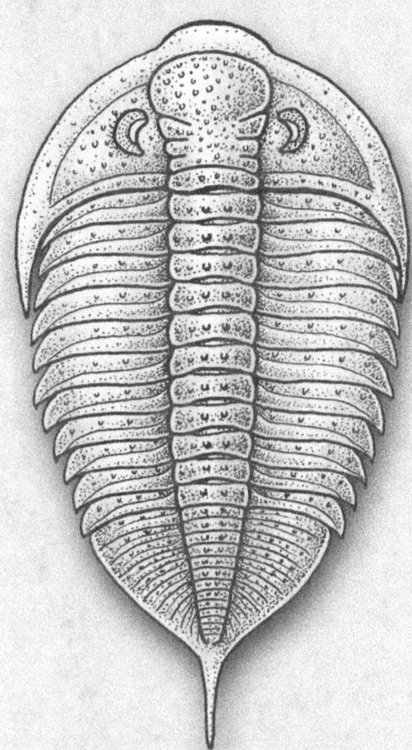

TRILOBITE - *Dalmanites*

CYSTOID - *Lepocrinites*

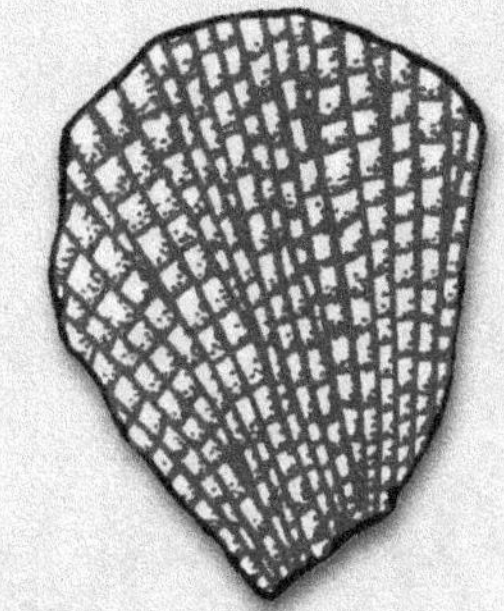

BRYOZOAN - Fenestella

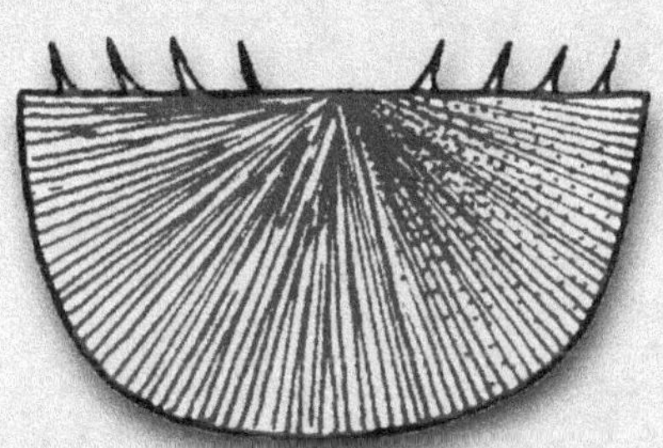

BRACHIOPOD - *Chonetes*

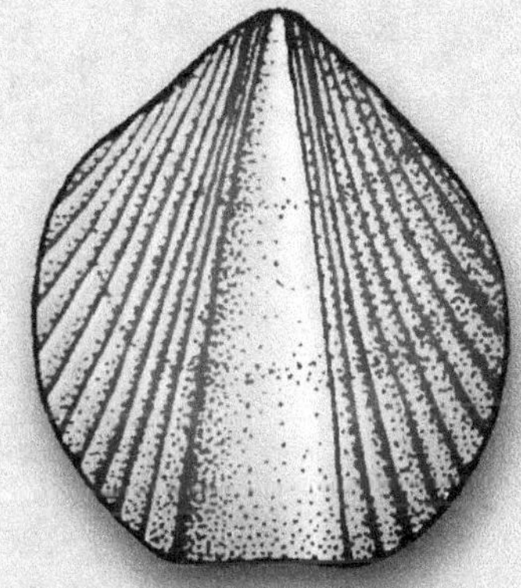

BRACHIOPOD - *Homeospira*

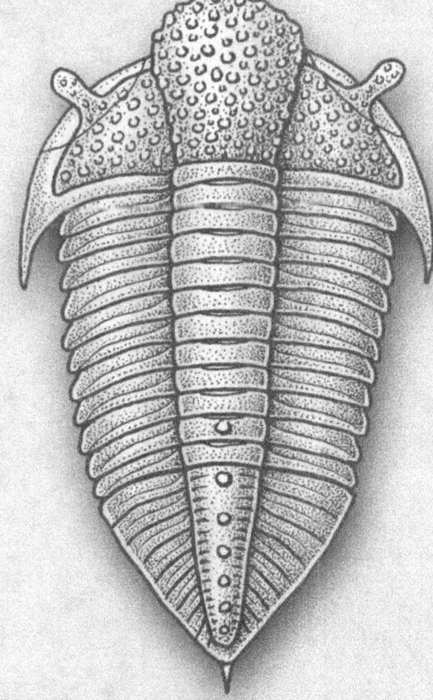

TRILOBITE - *Encrinurus*

BRACHIOPOD - *Cupularostrum*

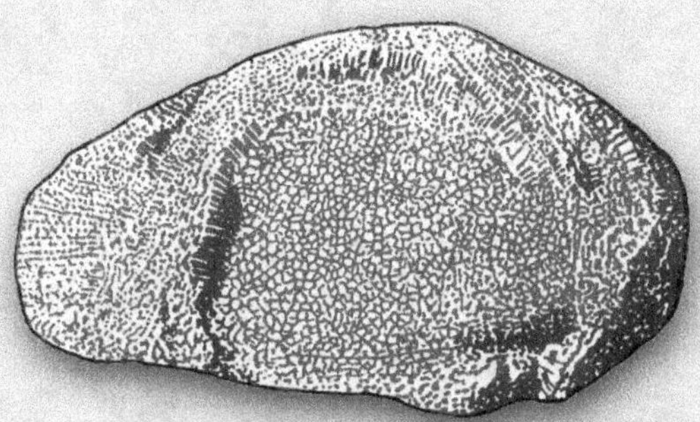

CORAL - *Favosites*

The Devonian Period in Virginia
(419-359 million years ago)

The Devonian Period has been called "The Age of Fishes." However, fish fossils of this age are comparatively scarce in Virginia. Invertebrate marine fossils, on the other hand, are abundant. Trilobites remained common, as did crinoids, corals, bivalves, gastropods, nautiloids and other shelled cephalopods, and bryozoans. Fossils of land plants from early forests that developed during the Devonian may be found.

As with the Ordovician Period in Virginia, the early Devonian is marked by considerable limestone accumulation. However, this gave way as the period progressed to mud and sand deposition as mountain building increased rates of erosion from land to sea. The results today are middle and upper Devonian sandstones and shales, some of which contain abundant fossils. The facing picture is based on specimens found in the Middle Devonian Mahantango Formation.

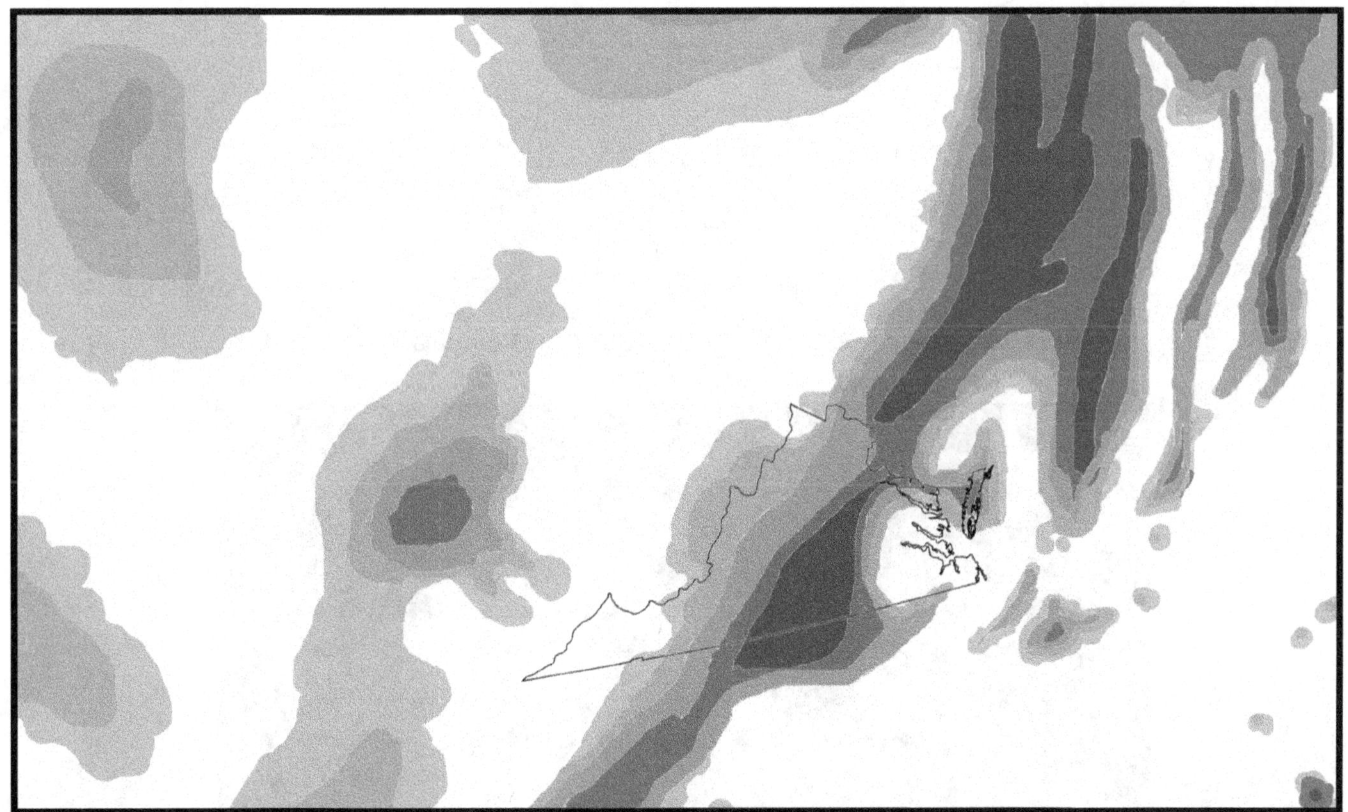

Elevation map of the region of Virginia (outline) during the Devonian Period, with white representing water and progressively darker shades of gray representing increasing land elevations. (After a map by Dr. Ron Blakey, Northern Arizona University.)

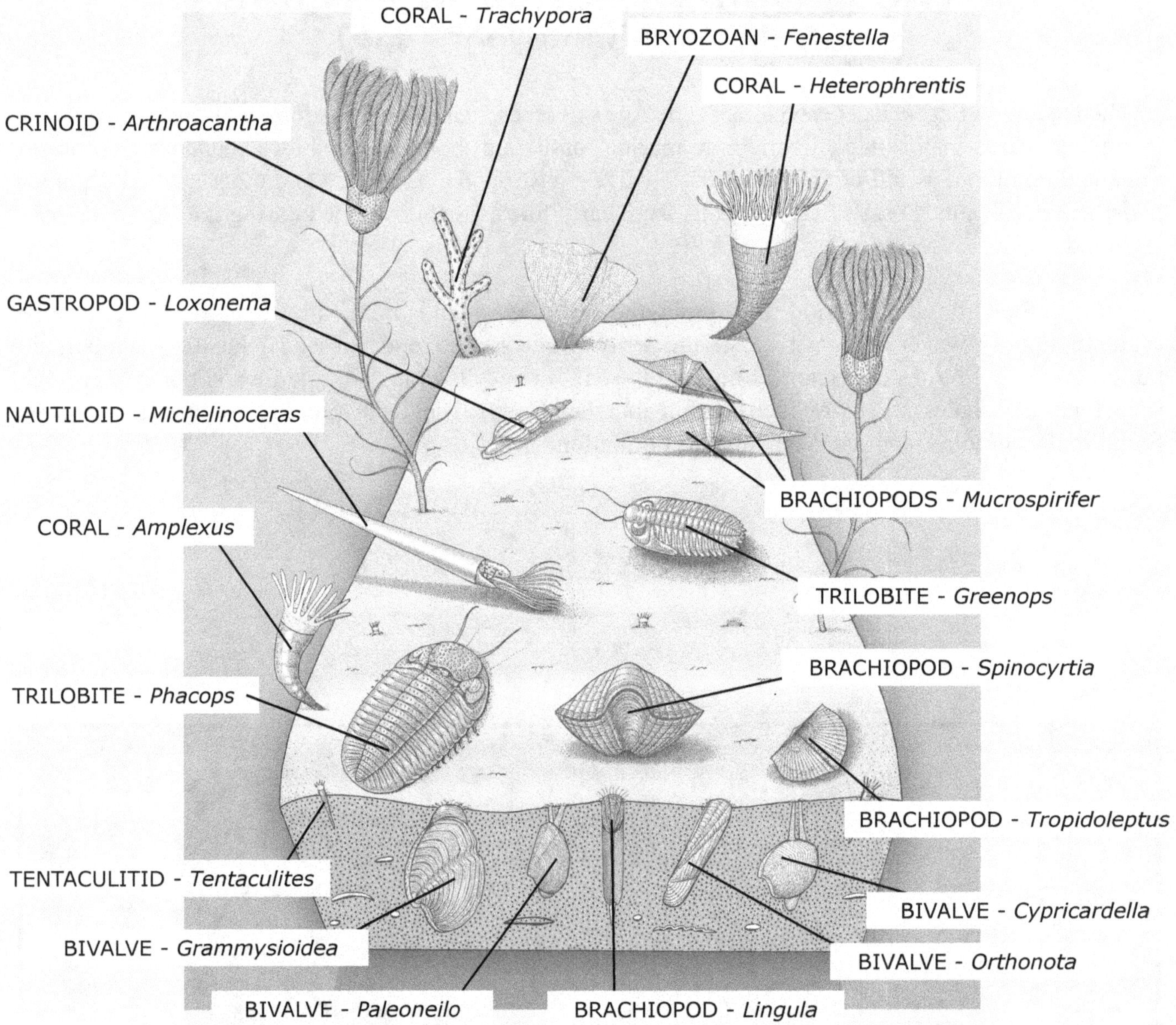

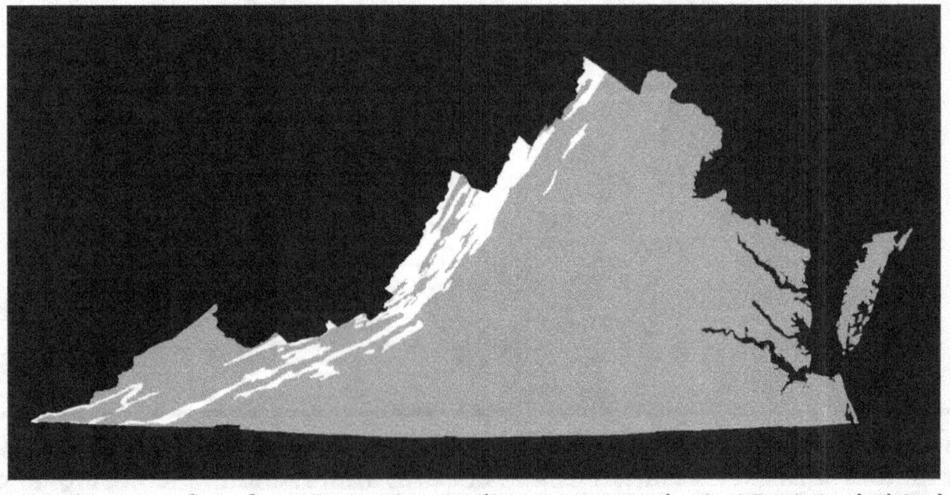

Distribution of surface Devonian sedimentary rocks in Virginia (white)

Devonian Fossils of Virginia

BIVALVE - *Platyostoma*

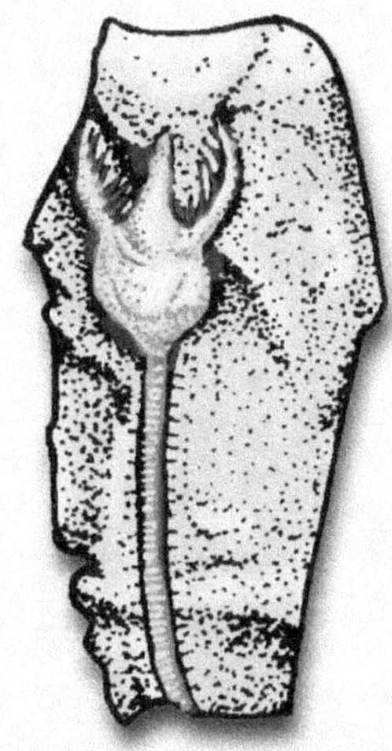

CRINOID - *Arthroacantha*

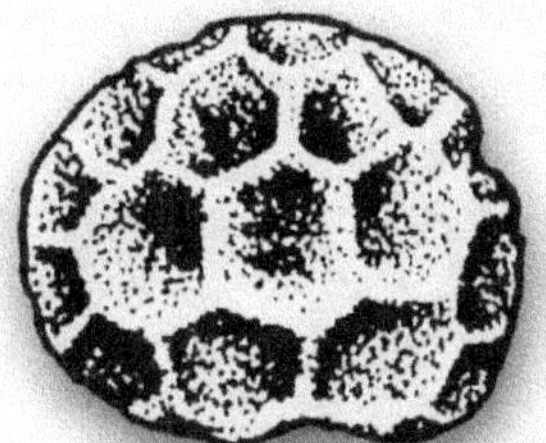

CORAL - *Pleurodictyum*

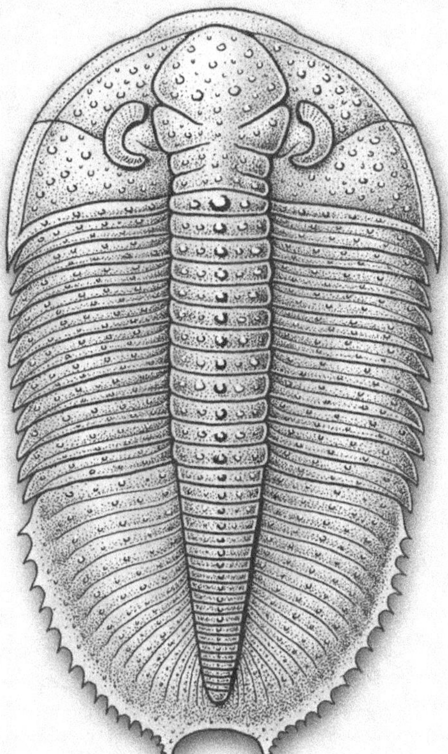

TRILOBITE - *Coronura*

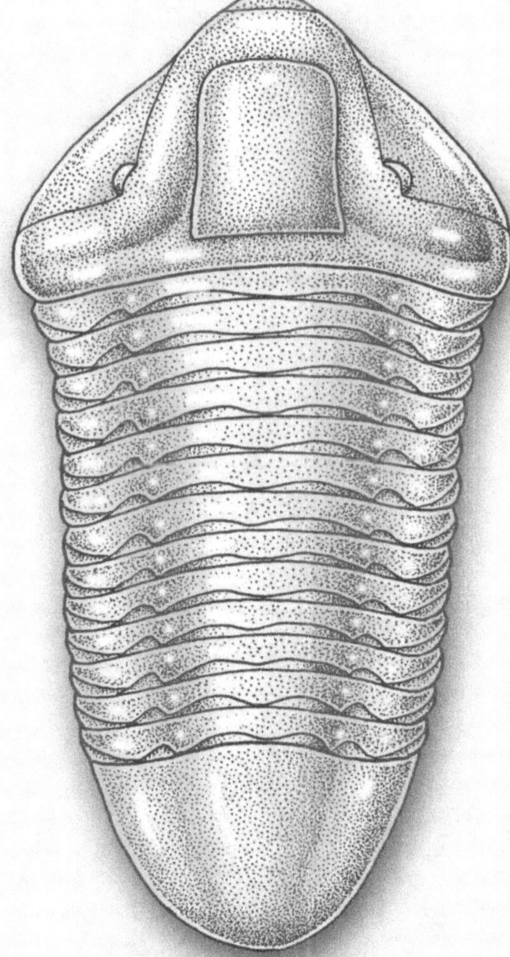

TRILOBITE - *Trimerus*

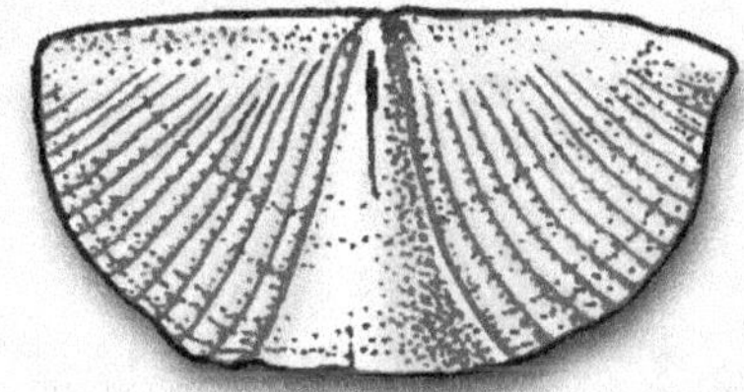

BRACHIOPOD - *Spinocyrtia*

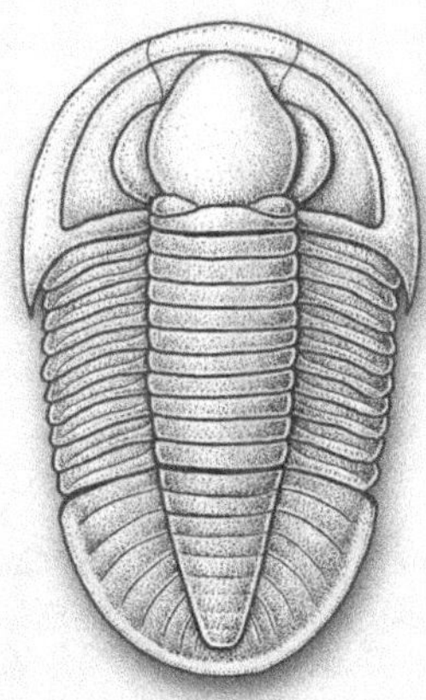

TRILOBITE - *Dechenella*

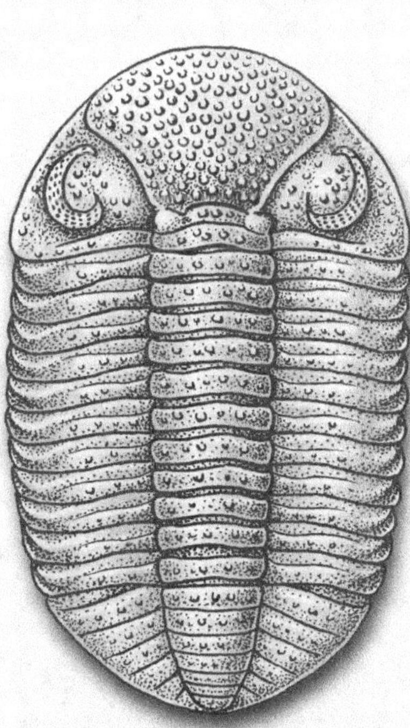

TRILOBITE - *Phacops*

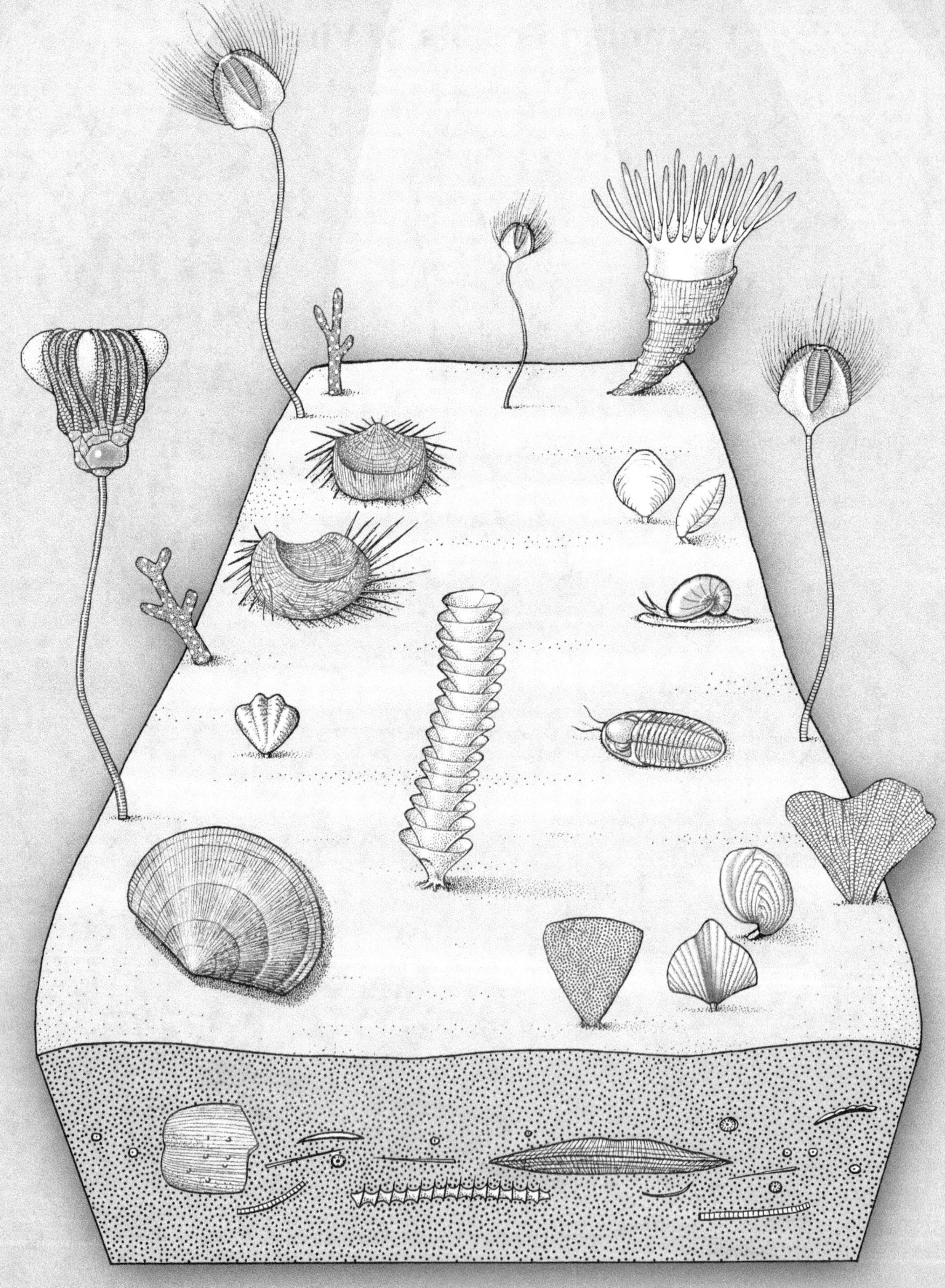

The Mississippian Period in Virginia
(359-323 million years ago)

The forests that became established in the Devonian Period became more extensive and varied during the Mississippian and important, well-preserved plant fossils from this period have been found in Virginia (e.g. in the Price Formation). Fossils of marine animals may be found in limestones – especially of crinoids and their relatives, the blastoids, both of which attained unprecedented abundance during the Mississippian. Also numerous are corals, brachiopods, bryozoans, and gastropods. Trilobites, on the other hand, are much less common in Mississippian rocks than previously.

The reconstruction opposite is based on the fossil assemblage that is found in limestones of Greenbrier (Middle Mississippian) age (345-326 mya).

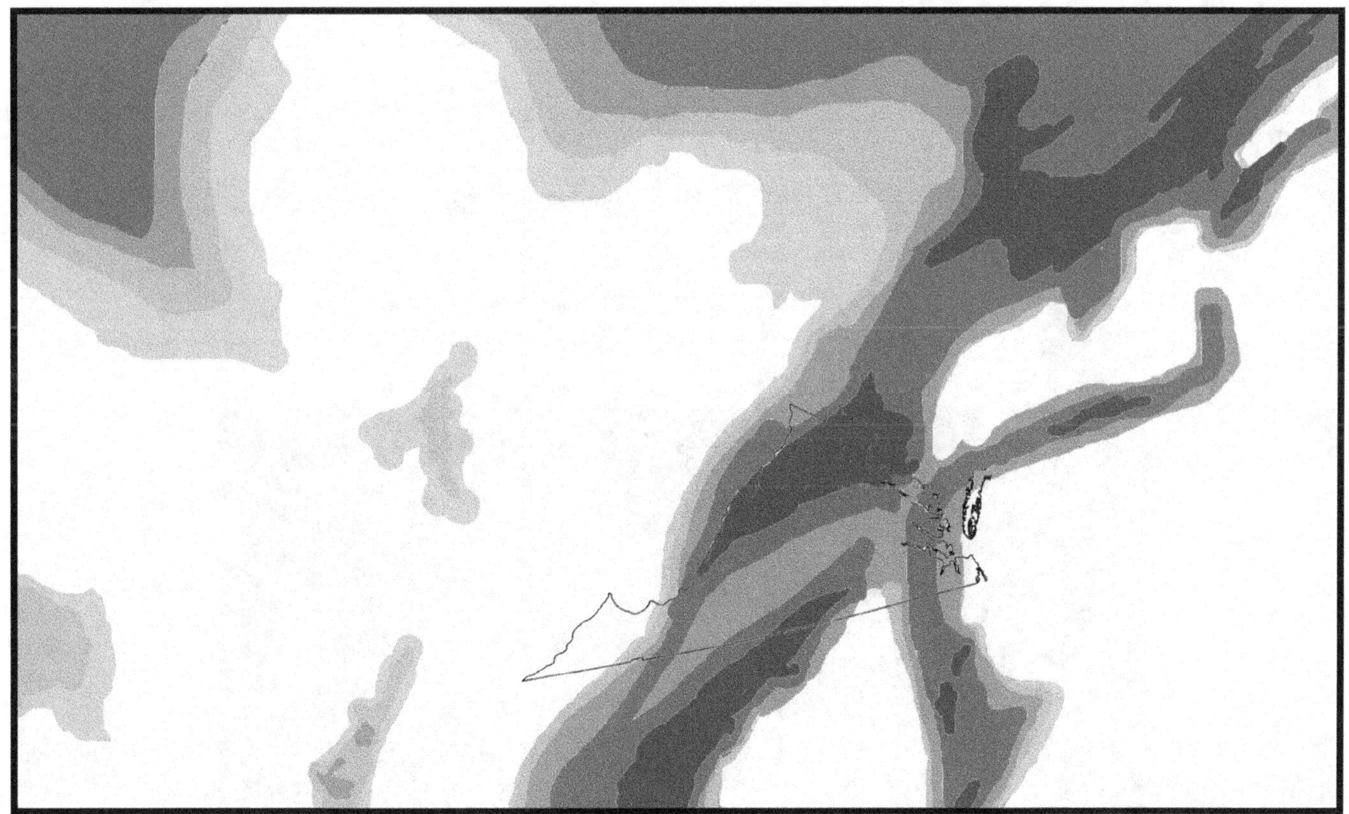

Elevation map of the region of Virginia (outline) during the Mississippian Period, with white representing water and progressively darker shades of gray representing increasing land elevations.
(After a map by Dr. Ron Blakey, Northern Arizona University.)

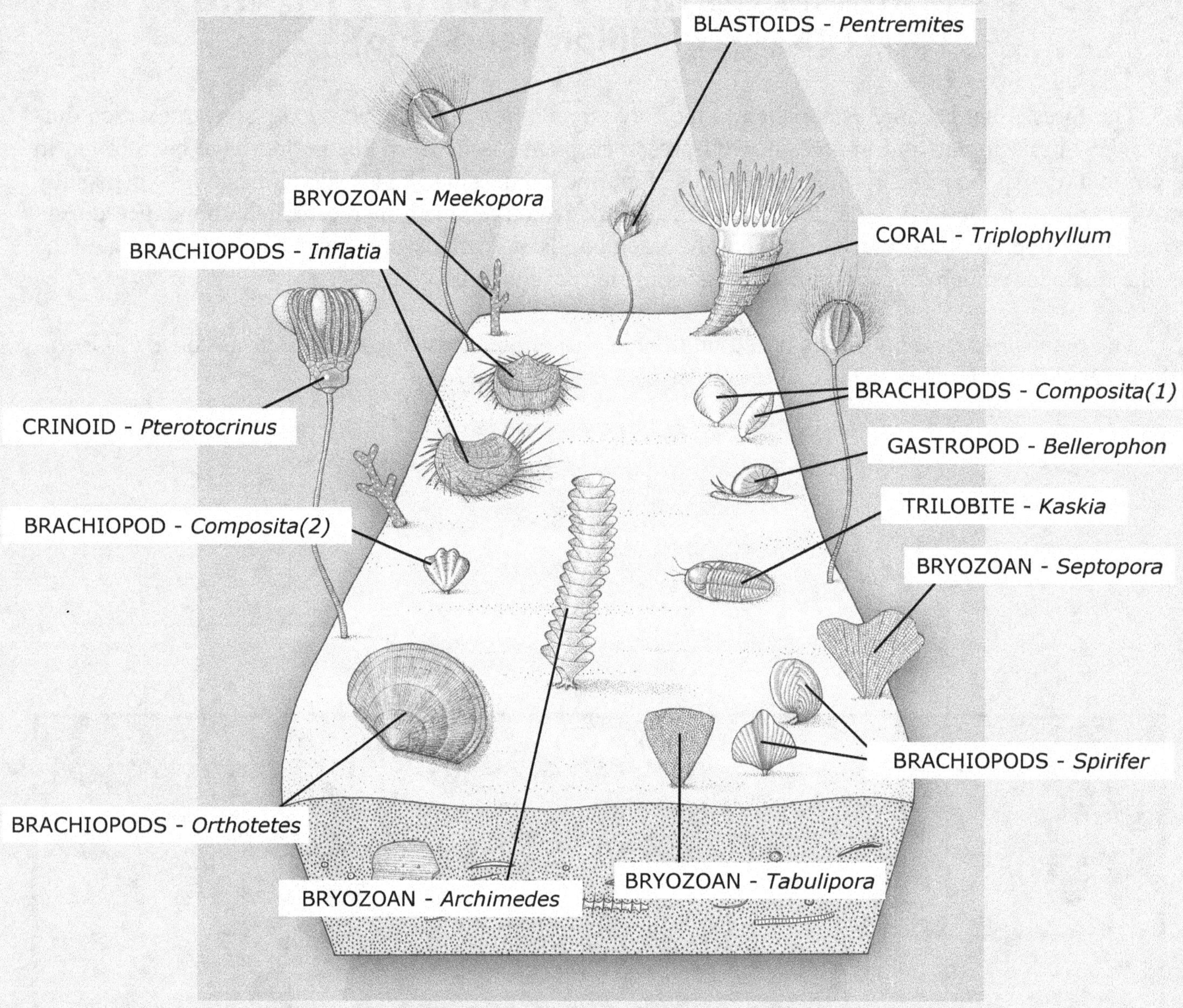

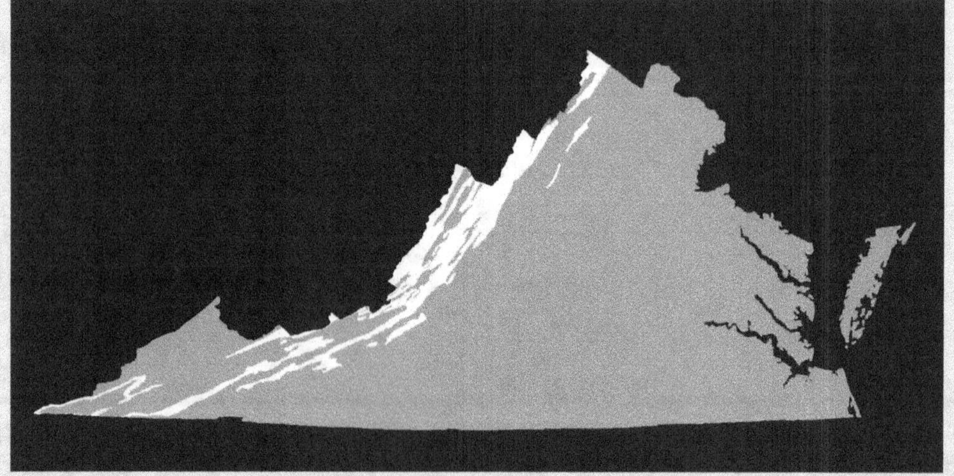

Distribution of surface Mississippian sedimentary rocks in Virginia (white)

Mississippian Fossils of Virginia

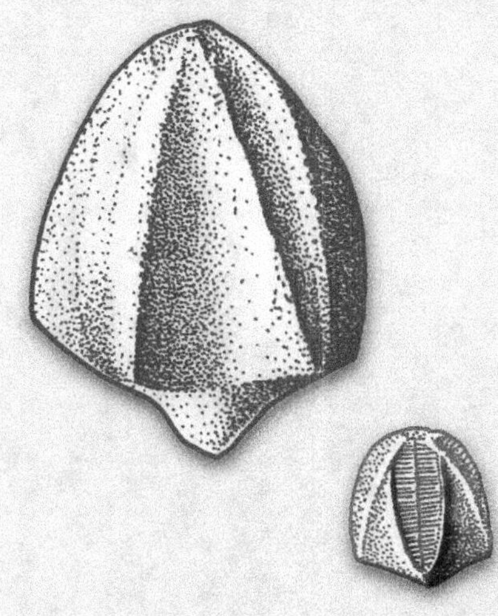

BLASTOIDS - *Pentremites*

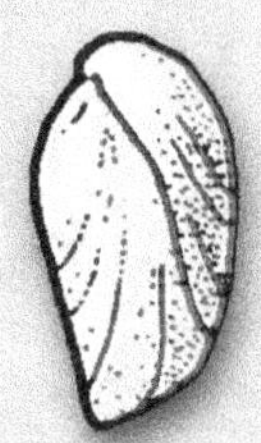

BRACHIOPODS - *Composita*

BRYOZOAN - *Septopora*

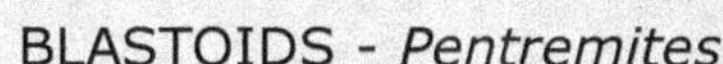

BIVALVE - *Posidonia*

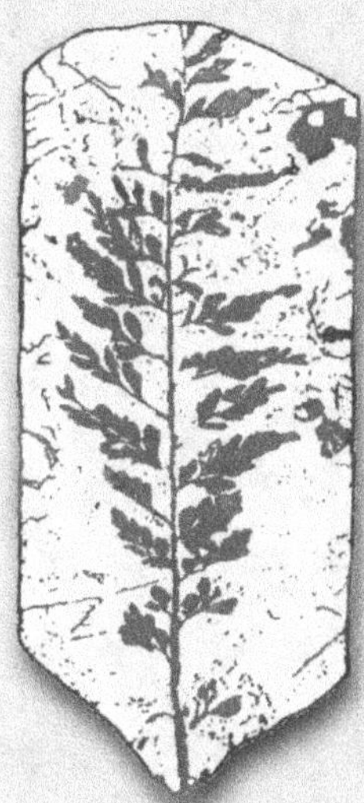

PLANT
Triphyllopteris

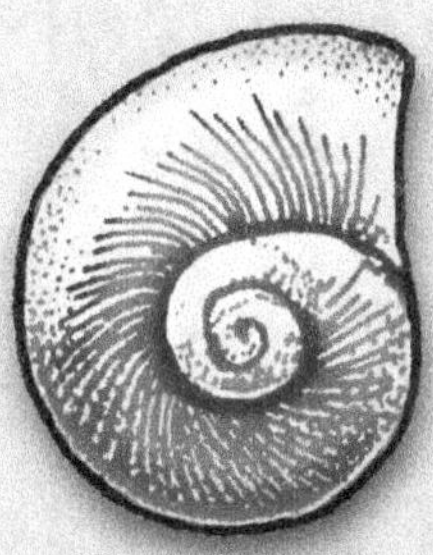

GASTROPOD - *Strophostylus*

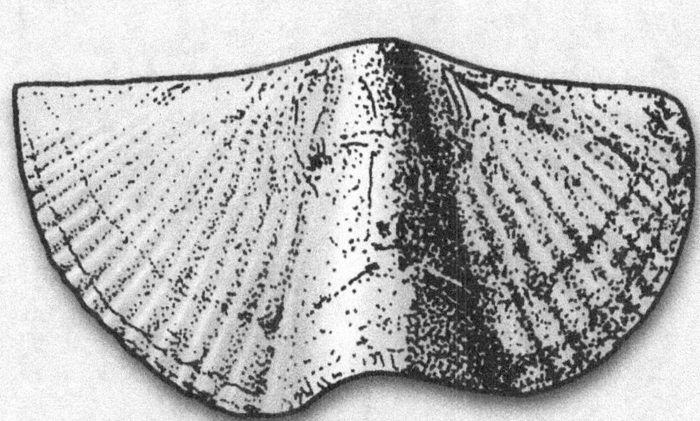

BRACHIOPOD - *Syringothyris*

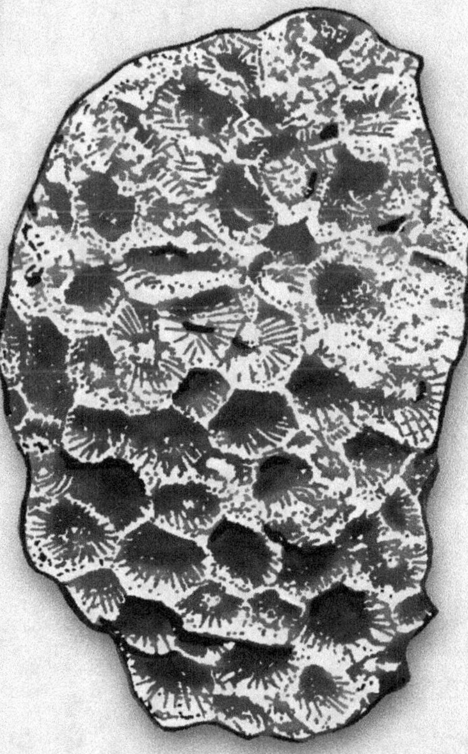

CORAL - *Lithostrotionella*

The Pennsylvanian Period in Virginia
(323-299 million years ago)

The forests of this period became so lush and extensive that their remains became economically important seams of coal when buried under sediments. This coal is an important fossil fuel that is mined in southwestern Virginia and neighboring states. The kinds of plants that flourished in these forests include horsetails (e.g. *Calamites*), seed ferns (e.g. *Neuropteris, Alethopteris*), and scale trees (e.g. *Lepidodendron*). Occasionally fossils of the animals that inhabited these forests are found, such as amphibians, fish, and various kinds of insects, centipedes, and millipedes. The remains of early reptiles and spiders have also been found.

At this time, the Earth's continents were coming together into one large land mass known as Pangaea. What would become Virginia was located near the center of this supercontinent and, for the first time in the Paleozoic Era, virtually no marine environments existed in the state. With the disappearance of large bodies of water, sedimentation decreased and Virginia has no fossil record of the following geologic period – the Permian. However, deposits in other states (e.g. West Virginia) show that many of the same kinds of plants and animals that thrived during the Pennsylvanian continued to live here during the Permian, though major changes were in store at the end of that period.

The scene opposite is based on fossils finds from Pennsylvanian coal-bearing strata in Virginia and adjoining states.

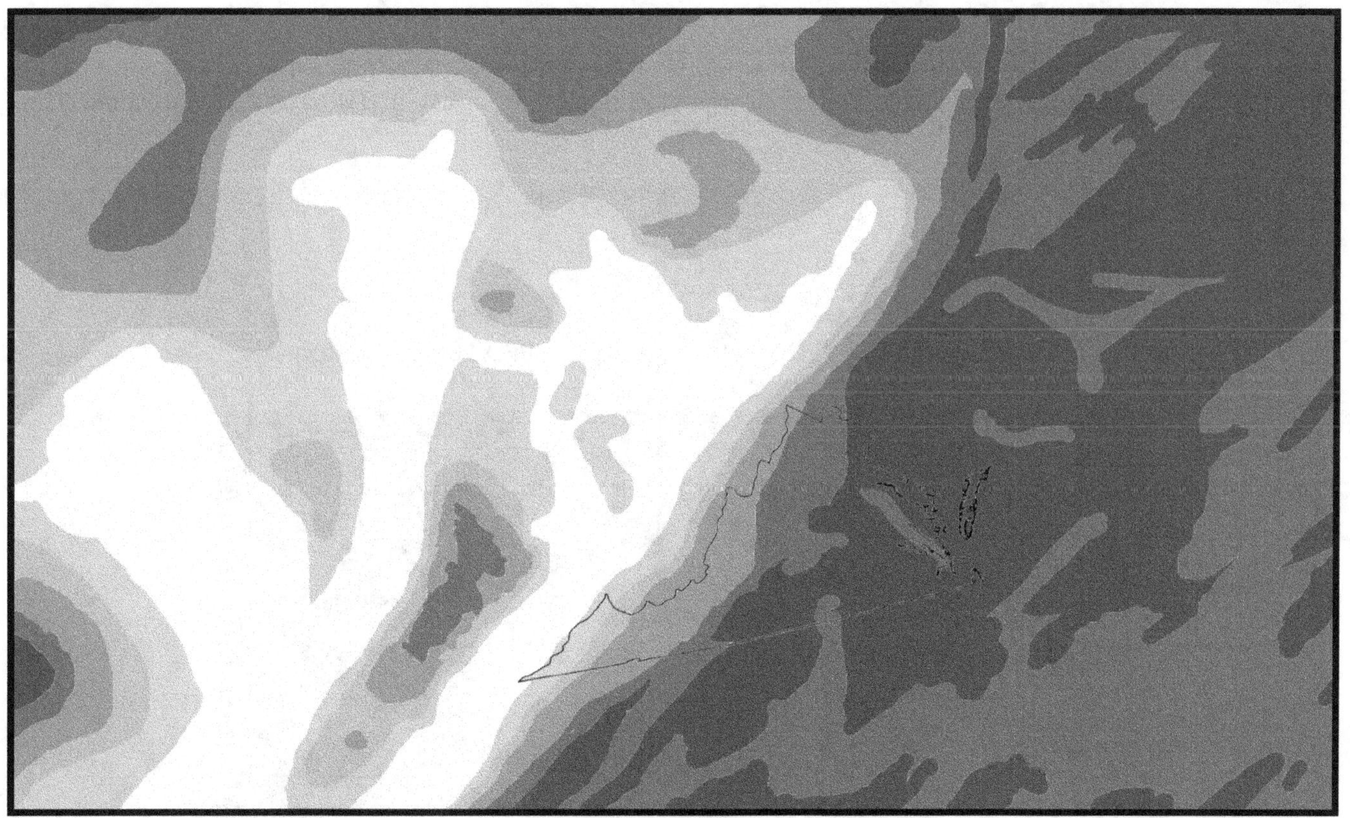

Elevation map of the region of Virginia (outline) during the Pennsylvanian Period, with white representing water and progressively darker shades of gray representing increasing land elevations. (After a map by Dr. Ron Blakey, Northern Arizona University.)

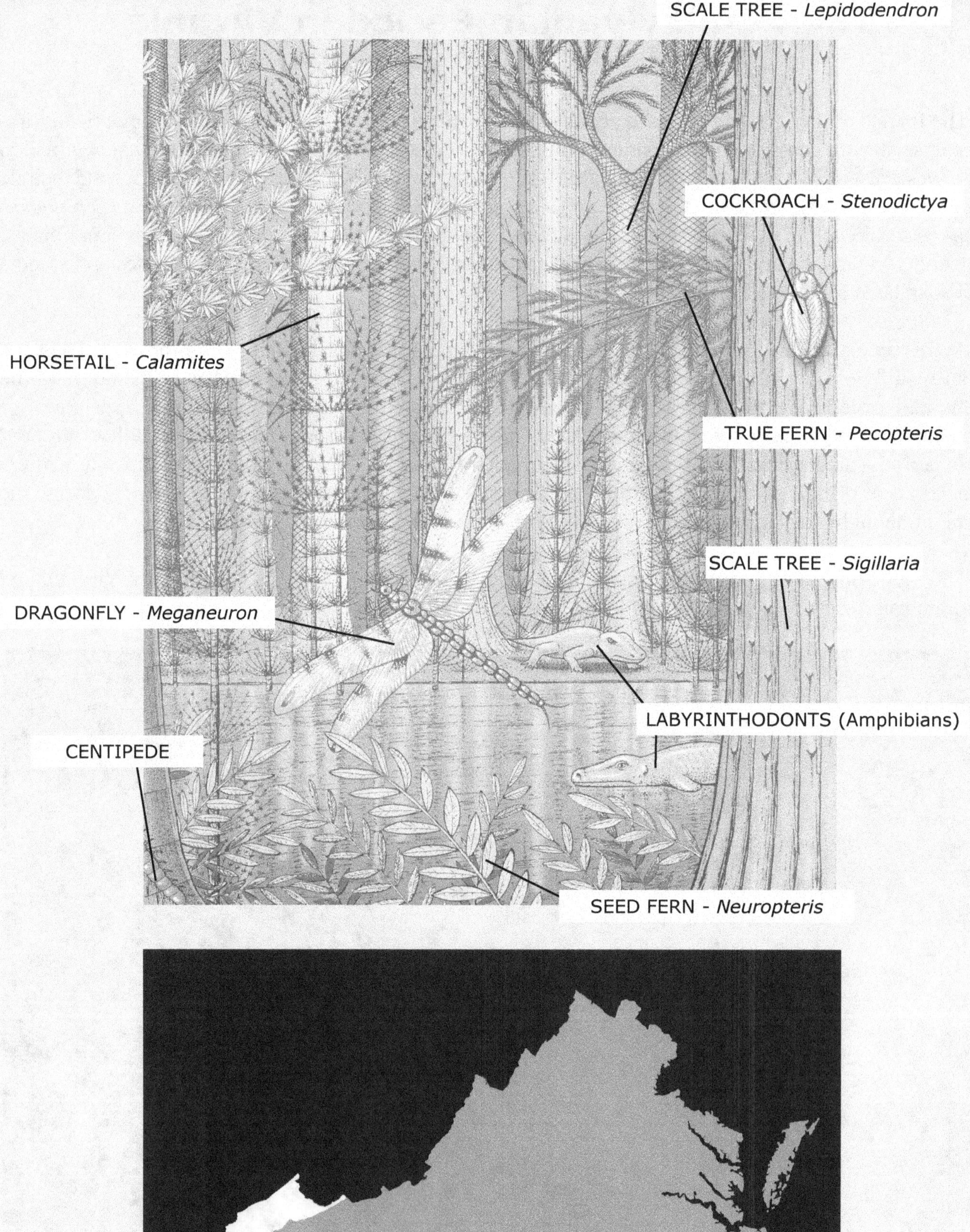

Distribution of surface Pennsylvanian sedimentary rocks in Virginia (white)

Pennsylvanian Fossils of Virginia

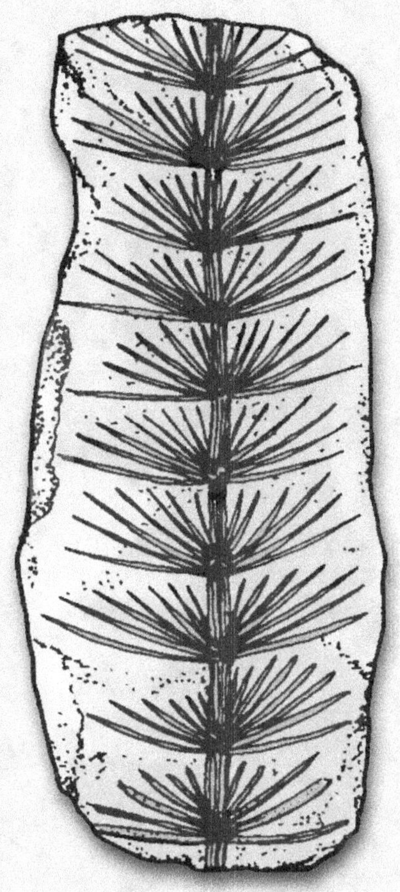

HORSETAIL - *Asterophyllites*

SEED FERN - *Alethopteris*

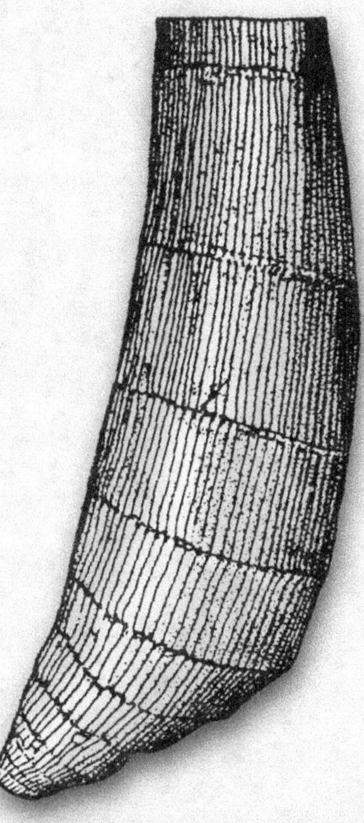

HORSETAIL - *Calamites*

SEED FERN - *Alethopteris*

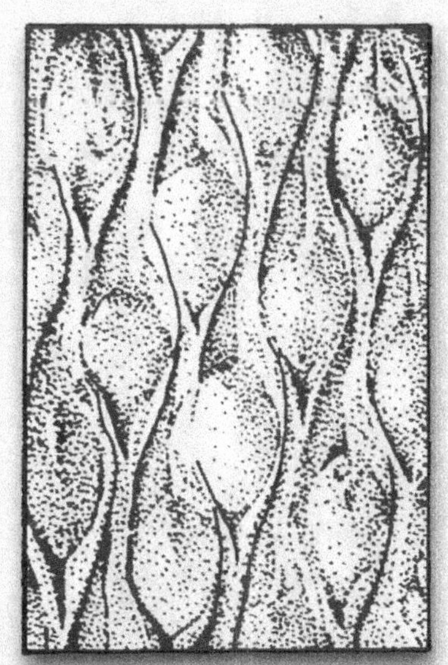

SCALE TREE - *Lepidodendron*

SEED FERN - *Mariopteris*

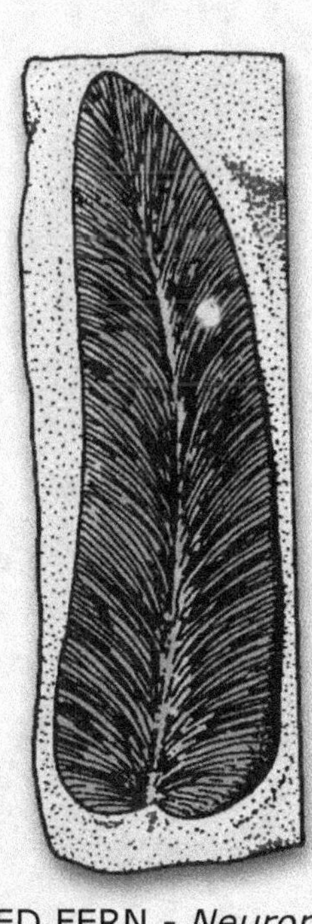

SEED FERN - *Neuropteris*

The Triassic-Jurassic Periods in Virginia
(252-145 million years ago)

The end of the Permian Period saw the extinction of many of the kinds of plants and animals that had been familiar residents of "Virginia" during the Paleozoic Era. Trilobites, blastoids, cystoids, and many other groups of marine invertebrates were gone forever. In fact, it is estimated that 96% of the marine species and 70% of the land species on Earth became extinct by the end of the Permian.

However, life bounced back, and the Triassic and early Jurassic sedimentary rocks of Virginia show that exciting new forms had moved into the area. These rocks were formed from deposits in and around lakes that existed in rift valleys that developed when the supercontinent Pangaea began to break up. One of these rift valleys continued to grow wider and wider and eventually became the Atlantic Ocean, dividing North America from other pieces of Pangaea that came to be known as Europe and Africa.

Dinosaur footprints and scarcer bones are among the most impressive Triassic and Jurassic fossils to be found in Virginia, though some other types are perhaps more unusual and scientifically important. Many fish fossils have been found, as well as the remains of plants (e.g. conifers, cycads, seed ferns, true ferns, and horsetails), insects (e.g. grasshoppers, beetles, flies, waterbugs, and cockroaches), and the long-necked aquatic reptile *Tanytrachelos*. The forests of this time became so luxuriant that coal deposits formed which have been commercially mined.

The scene opposite is based on fossil discoveries in the Triassic to Jurassic rocks of southern Virginia and northern North Carolina.

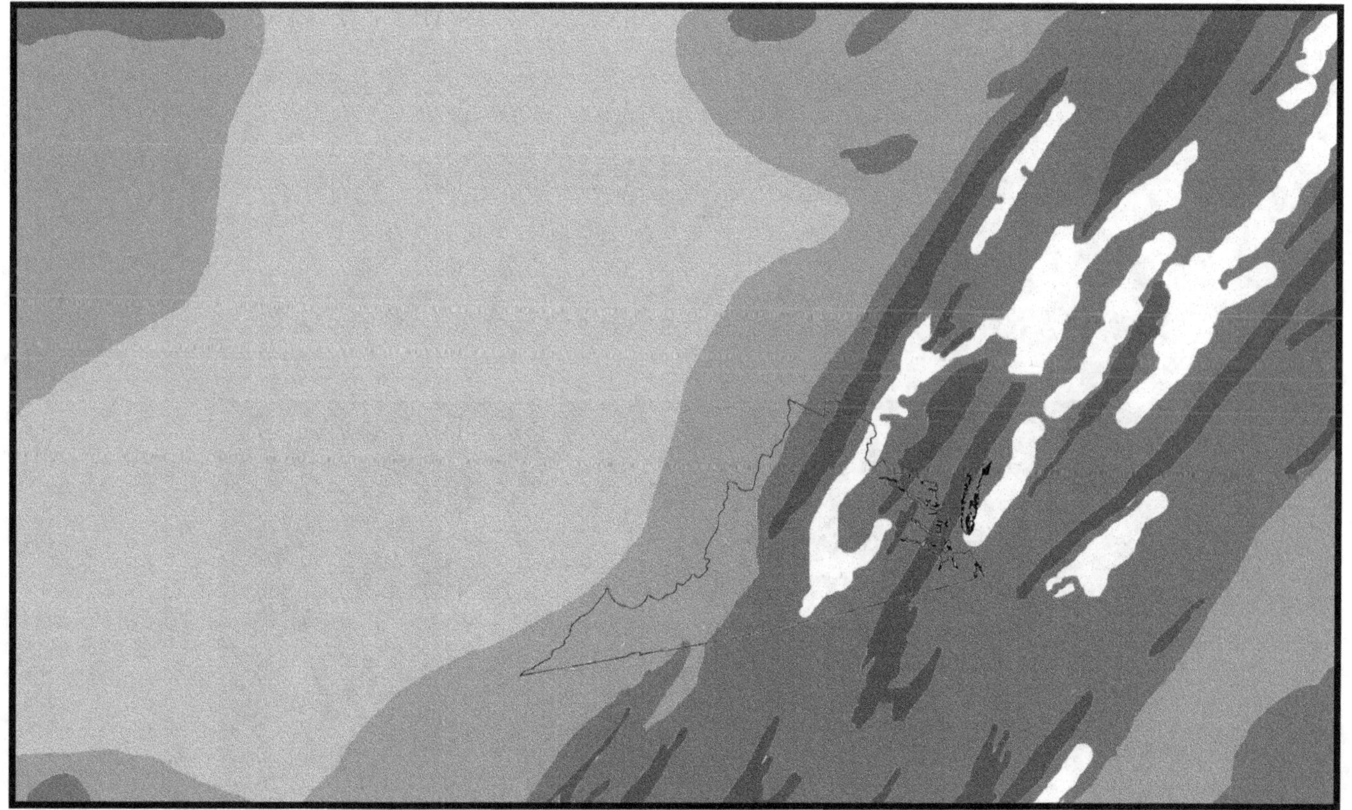

Elevation map of the region of Virginia (outline) during the Triassic Period, with white representing water and progressively darker shades of gray representing increasing land elevations. (After a map by Dr. Ron Blakey, Northern Arizona University.)

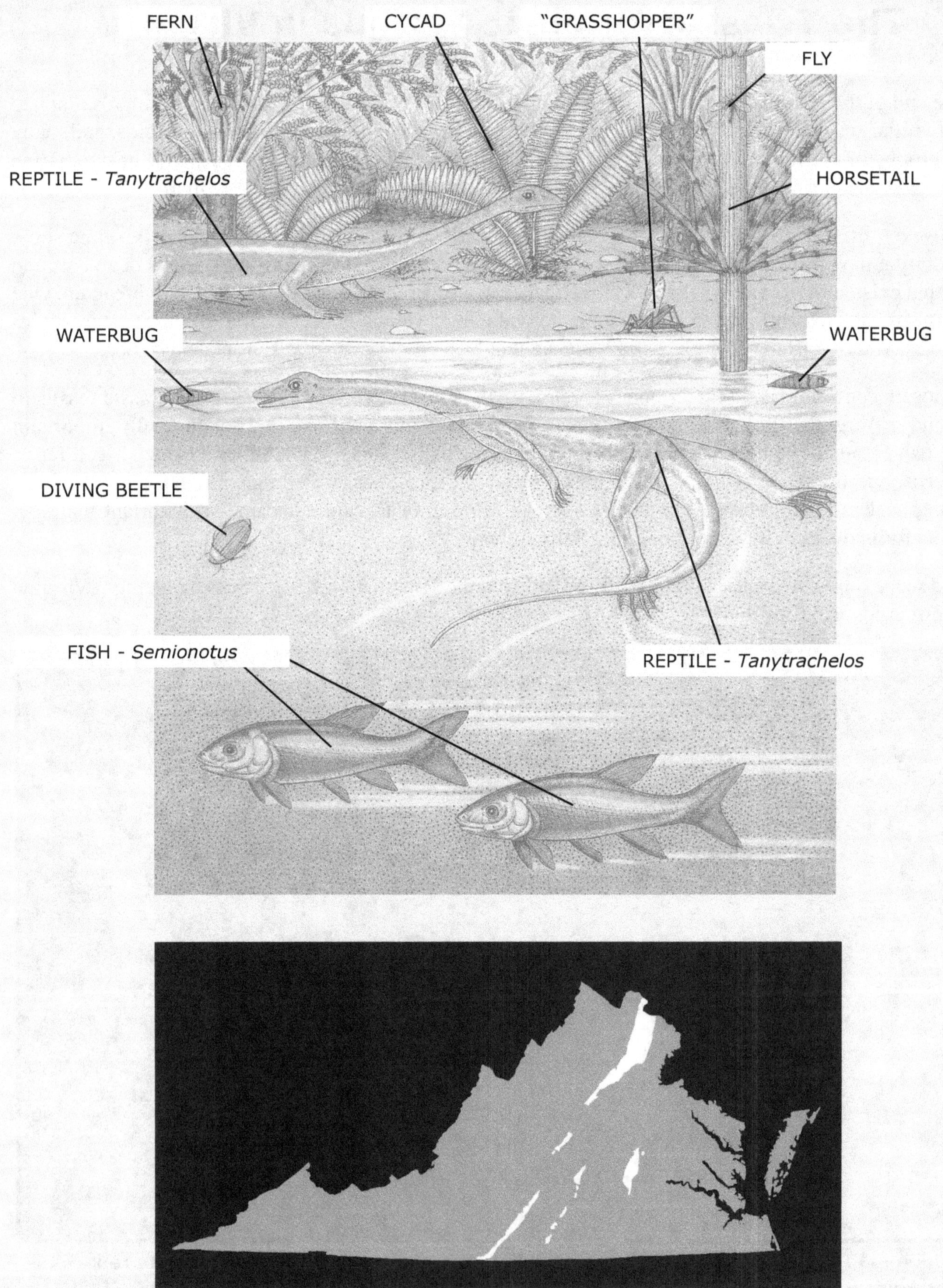

Distribution of surface Triassic and Jurassic sedimentary rocks in Virginia (white)

Triassic-Jurassic Fossils of Virginia

HORSETAIL

FISH - *Semionotis*

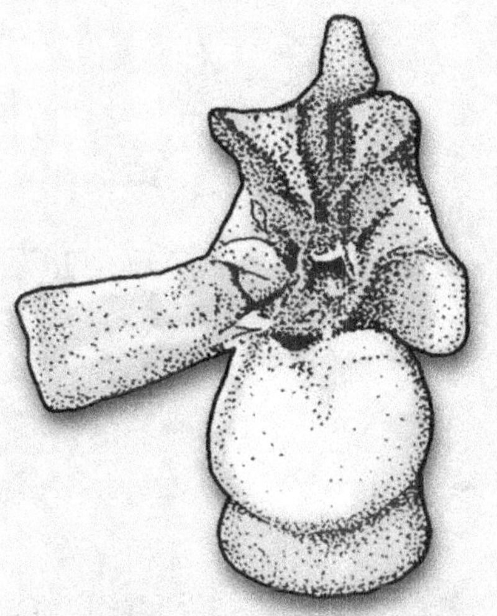

PHYTOSAUR - *Rutiodon* (vertebra)

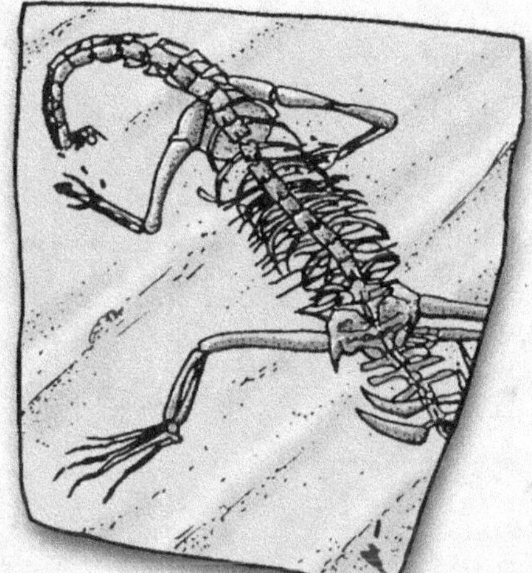

REPTILE - *Tanytrachelos*

FLY

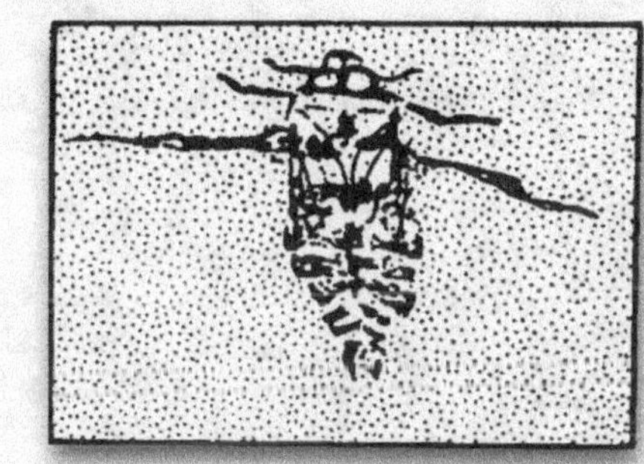

WATERBUG - *Triassonepa*

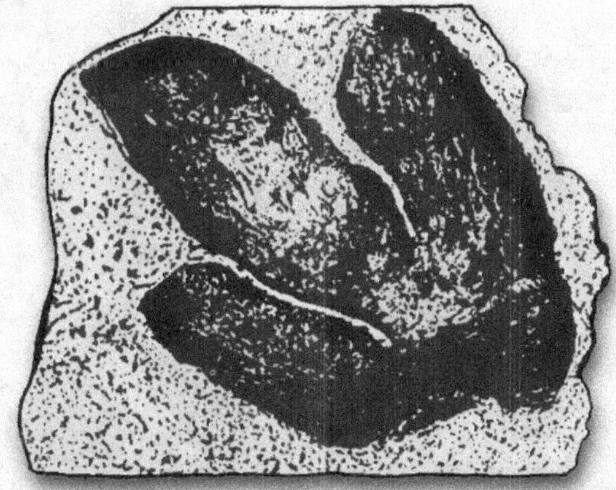

DINOSAUR TRACK - *Eubrontes*

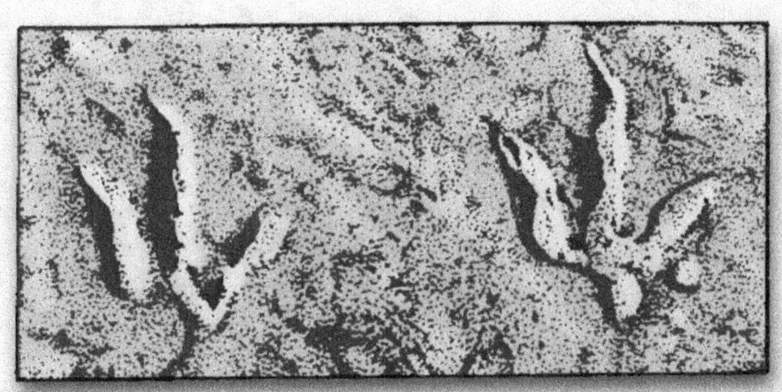

DINOSAUR TRACKS - *Grallator*

The Cretaceous Period in Virginia
(145-66 million years ago)

Though Virginia was certainly inhabited by a rich array of dinosaurs and other large reptiles during the Cretaceous, its fossils are primarily restricted to a rich variety of land plants, including cycads, seed ferns, redwoods (*Sequoia*), horsetails, and conifers. Some specimens of freshwater crustaceans and fish have been found, but one must look to the neighboring states of Maryland and North Carolina for fossils of large reptiles, and Virginia's Cretaceous marine fossils are too deeply buried under younger sediments to be collected.

As at the end of the Permian Period, the end of the Cretaceous saw the extinction of many forms of life, with as many as 75% of all living species ceasing to exist.

The drawing to the left shows many of the kinds of plants that left fossil remains in Virginia as well as the dinosaurs whose bones are found in neighboring states and may be presumed to have lived in Virginia as well.

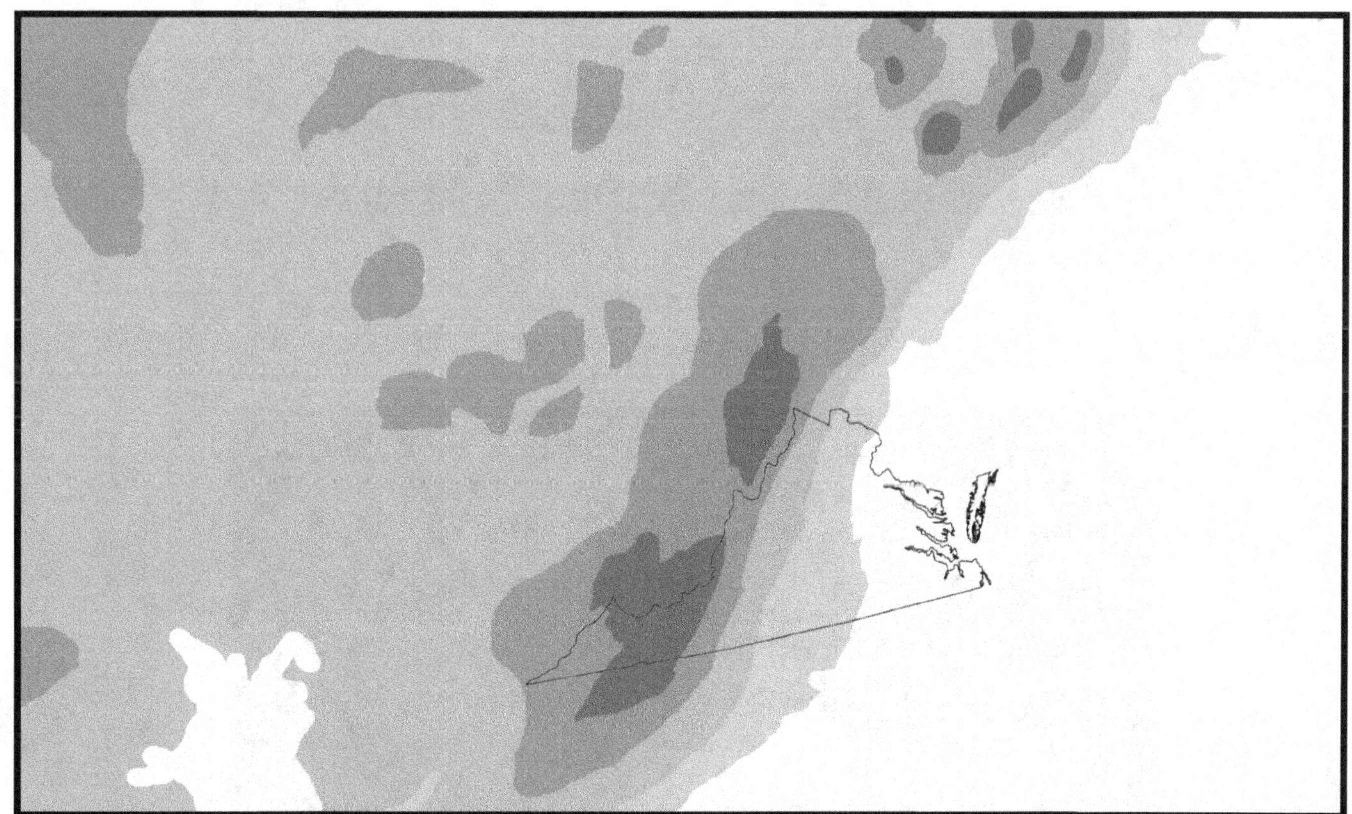

Elevation map of the region of Virginia (outline) during the Cretaceous Period, with white representing water and progressively darker shades of gray representing increasing land elevations. (After a map by Dr. Ron Blakey, Northern Arizona University.)

Distribution of surface Cretaceous sedimentary rocks in Virginia (white)

Cretaceous Fossils of Virginia

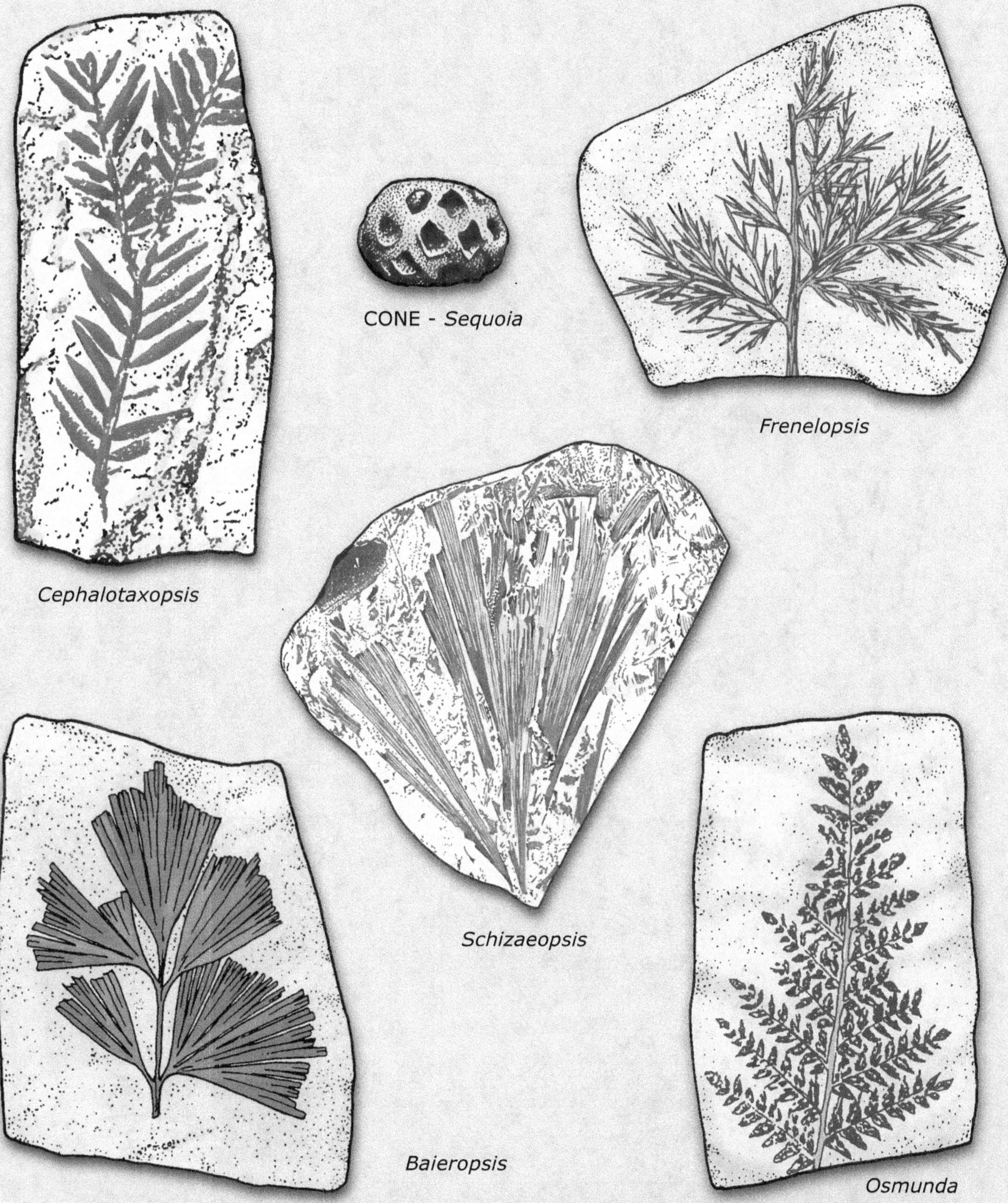

The Paleocene Epoch in Virginia
(66-56 million years ago)

Nearly all of the Mesozoic fossils to be found in Virginia are of land-based or freshwater plants and animals. However, most of the state's Cenozoic fossil record consists of marine specimens, and these comprise one of the best-preserved and most extensive fossil assemblages in the world.

Paleocene sediments contain an abundance of bones and teeth of aquatic turtles, crocodiles, sharks, rays, and other fish. Unlike later epochs of the Tertiary, these deposits have no whale fossils as these animals had not yet evolved. There are also abundant and distinctive invertebrate fossils, including oysters and many other bivalves, gastropods (e.g. *Turritella*), corals, and many more.

The reconstruction on the facing page is based on the fossils collected in the Upper Paleocene Aquia Formation of Virginia.

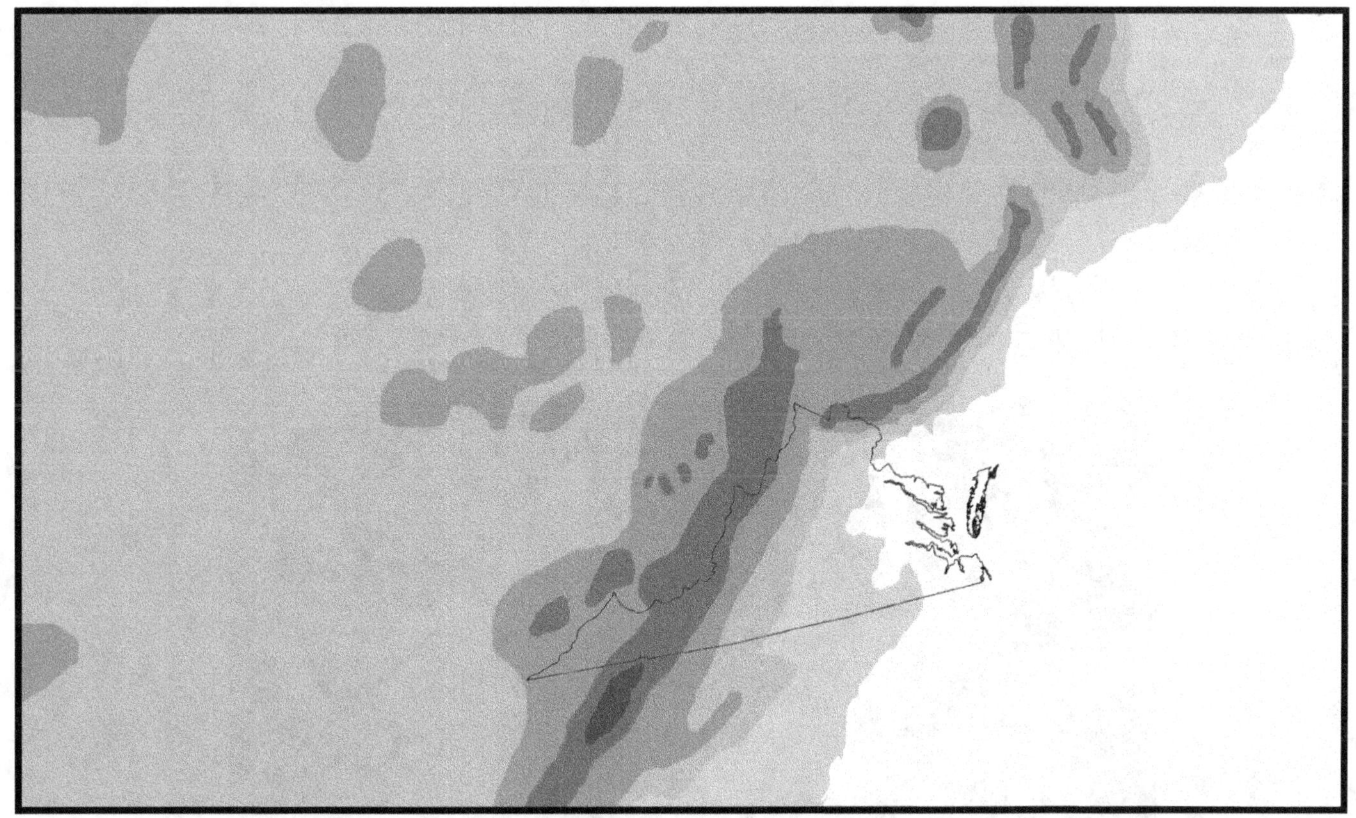

Elevation map of the region of Virginia (outline) during the Paleocene Epoch, with white representing water and progressively darker shades of gray representing increasing land elevations. (After a map by Dr. Ron Blakey, Northern Arizona University.)

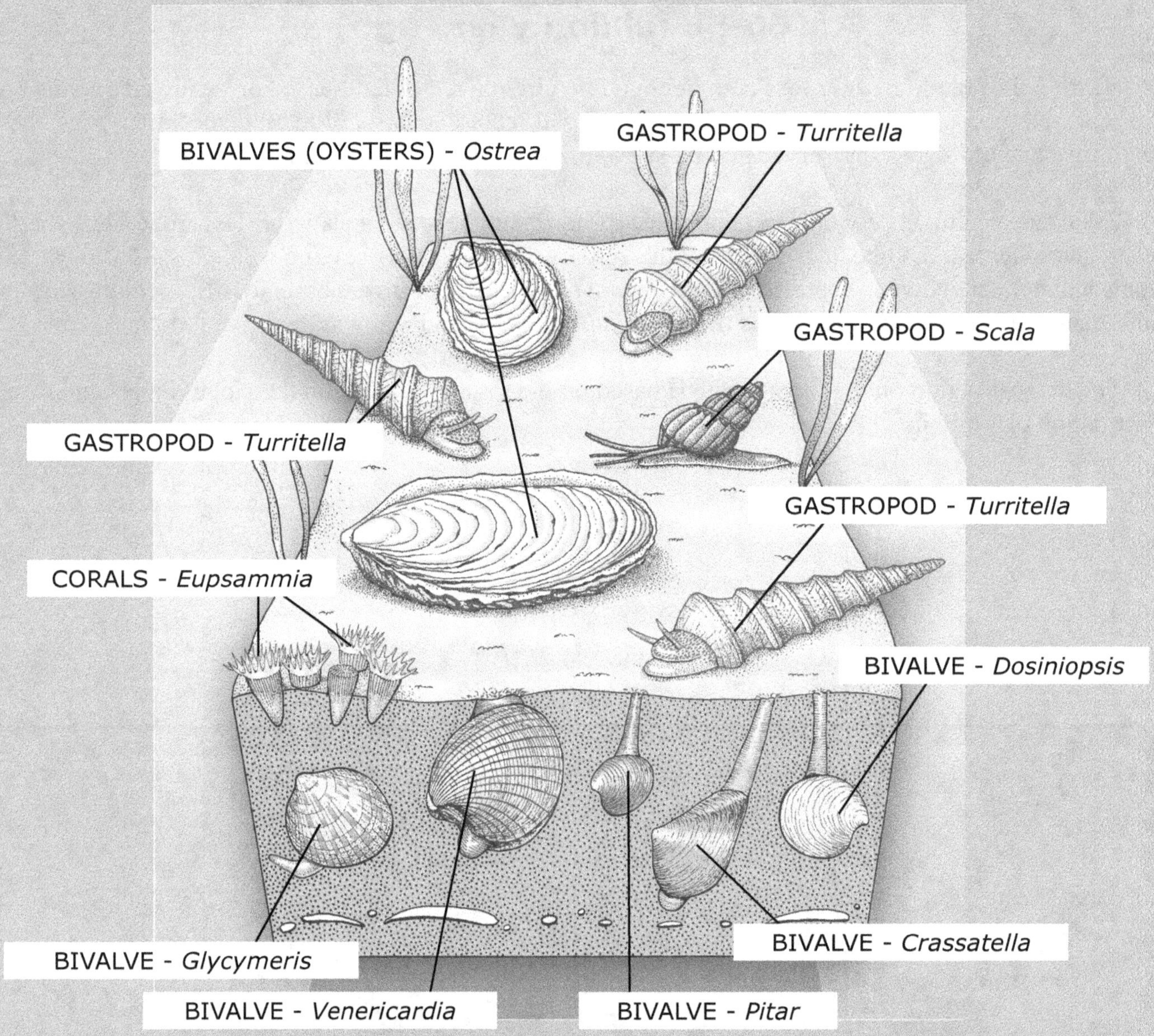

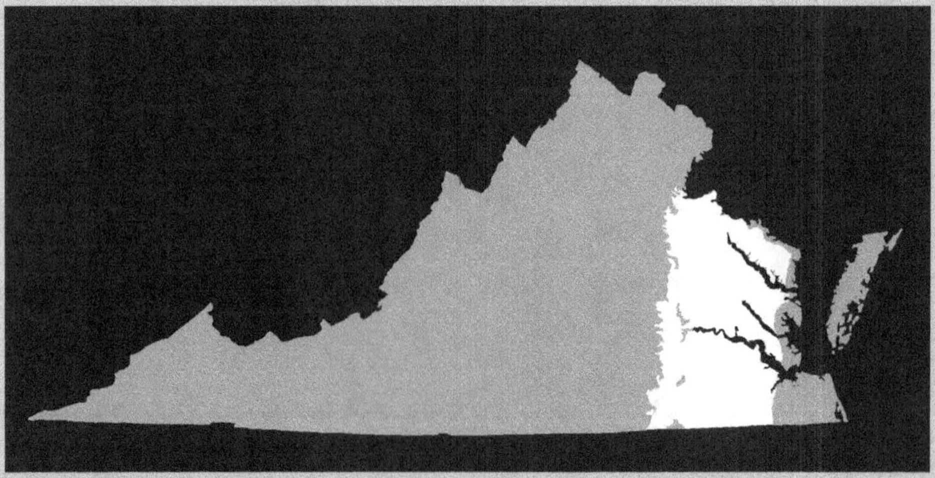

Distribution of surface Tertiary sedimentary rocks and deposits in Virginia (white)

Paleocene Fossils of Virginia

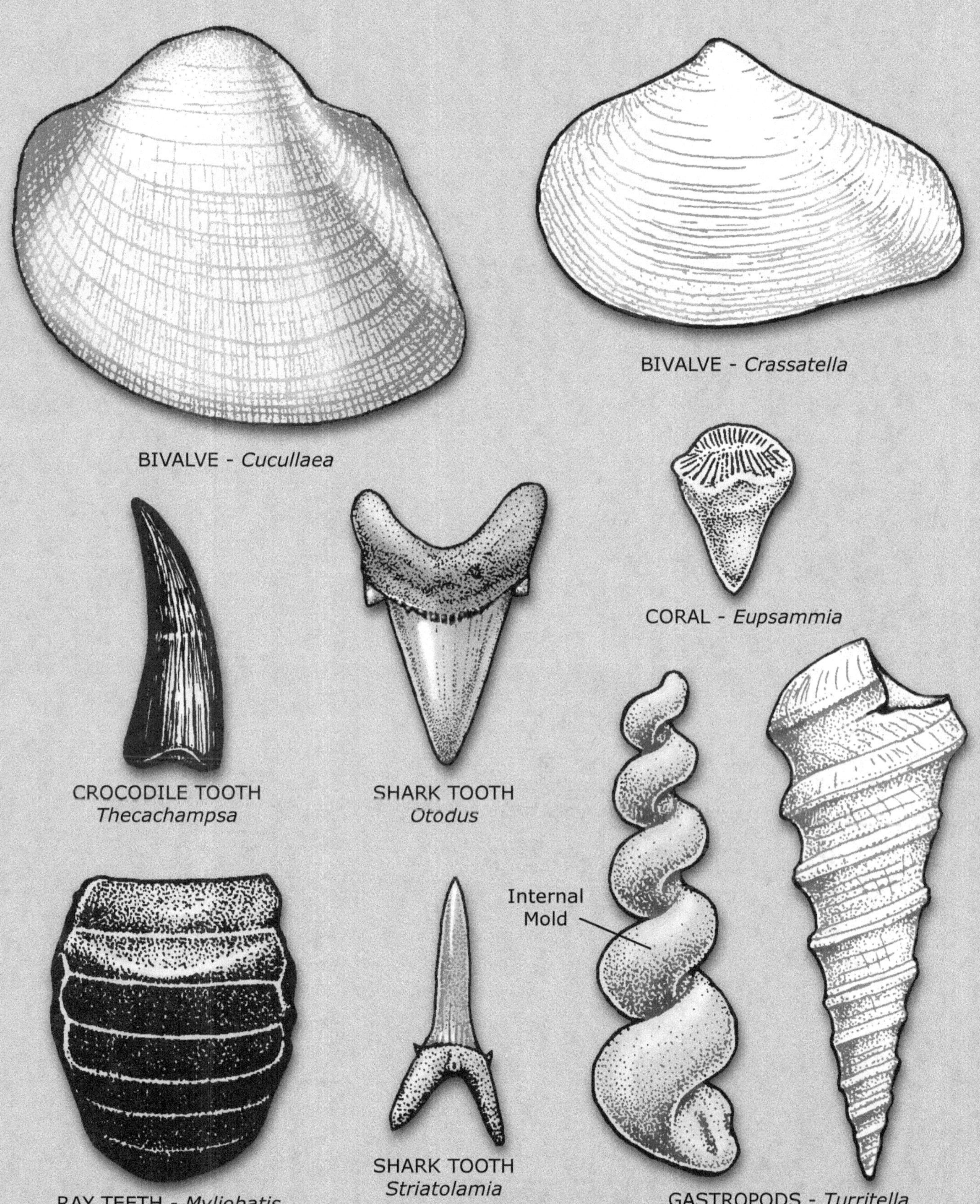

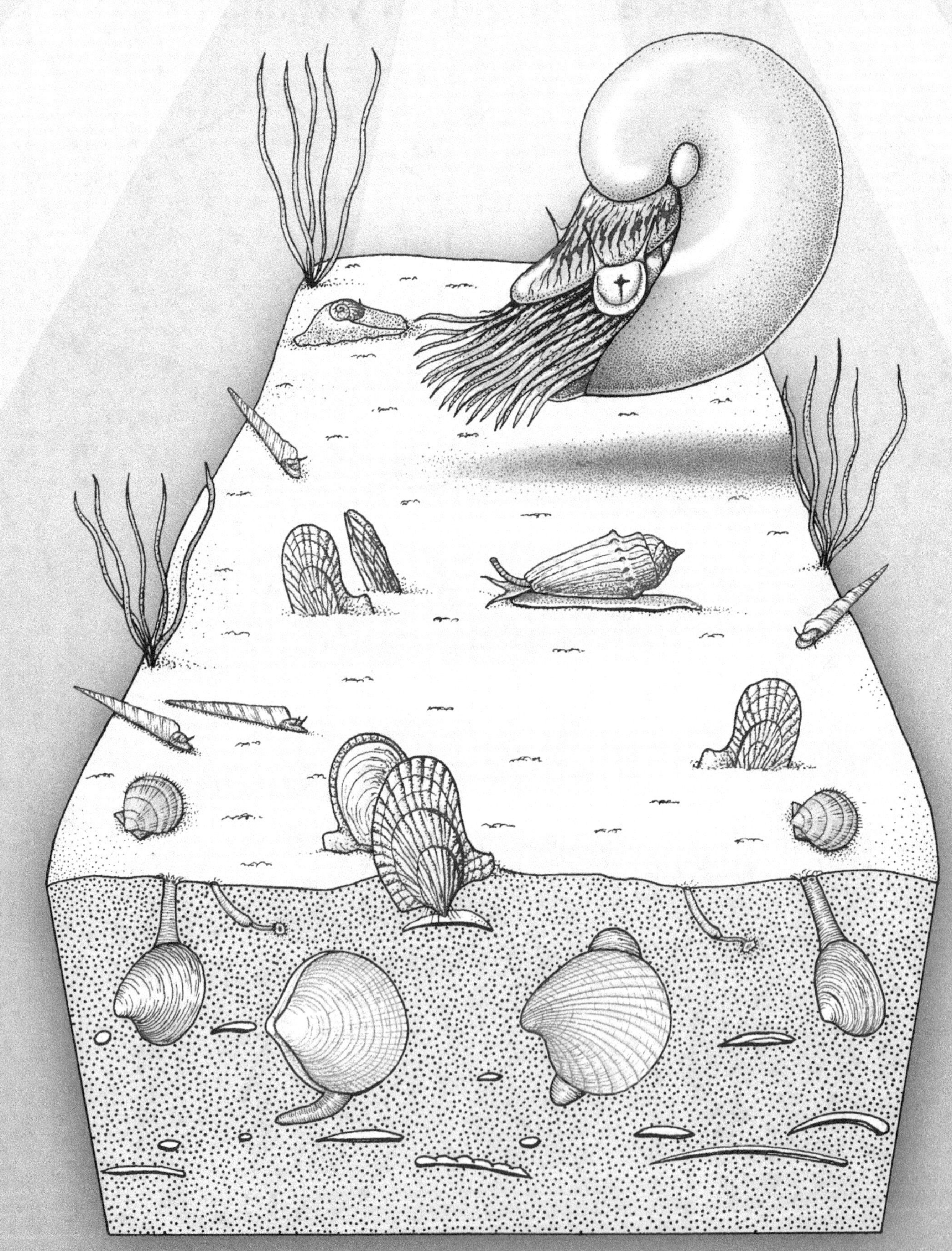

The Eocene Epoch in Virginia
(56-33.9 million years ago)

Virginia's Eocene fossils are similar to but less abundant than those found in Paleocene sediments. Bivalves and gastropods are the most common larger fossils – one species of clam (*Venericardia*) comprises up to 95% of the fossil content of some beds. Shark, ray, and other fish teeth are also found.

A rare find in Eocene deposits is the shell of the nautiloid *Hercoglossa*. Shelled cephalopods had been extremely common in the Paleozoic and Mesozoic Eras but were almost completely killed off during the Cretaceous extinction.

Fossils of the succeeding epoch of Earth history, the Oligocene, are all but nonexistent in Virginia. The few that have been found are mollusks similar to Miocene species.

The re-creation to the left is based on the fossil record in the Lower Eocene Nanjemoy Formation of eastern Virginia.

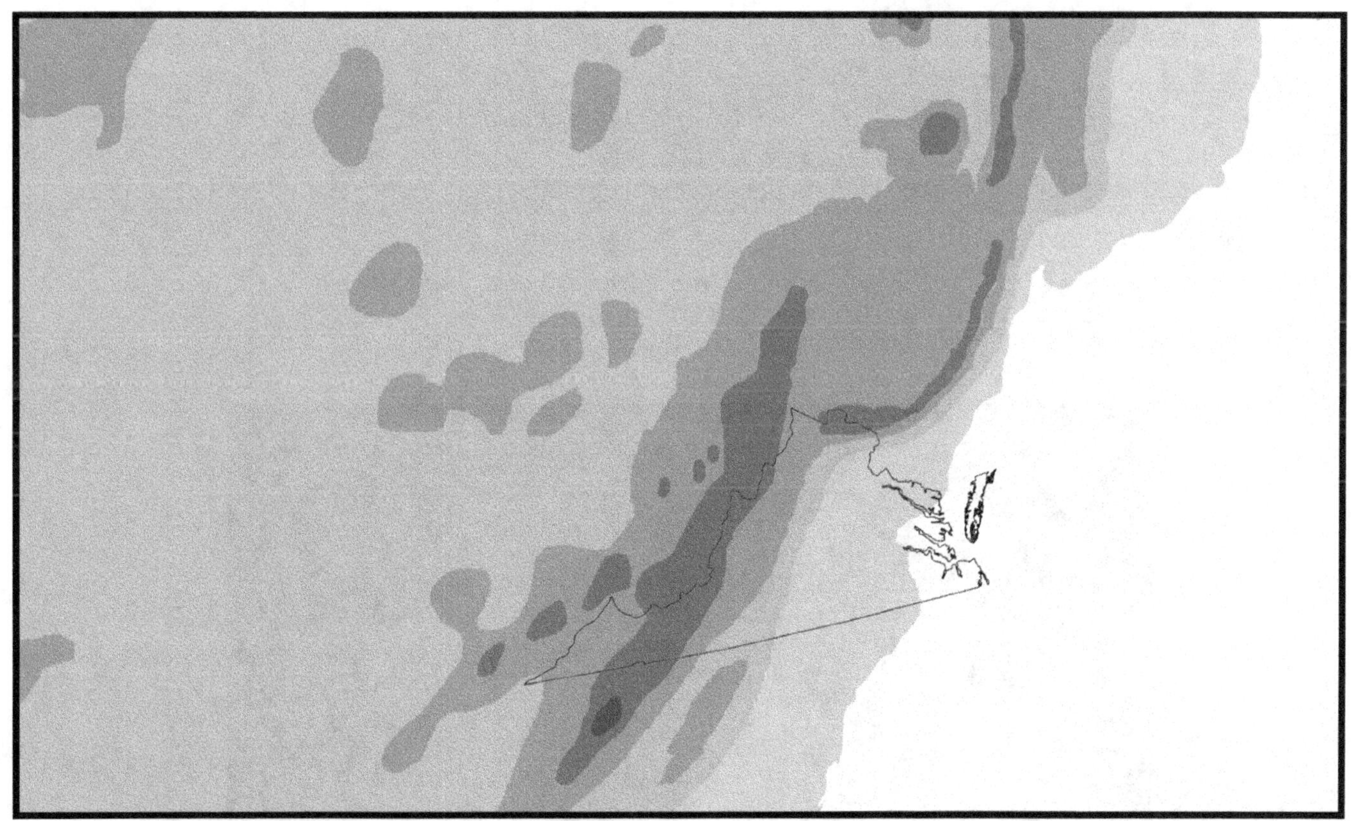

Elevation map of the region of Virginia (outline) during the Eocene Epoch, with white representing water and progressively darker shades of gray representing increasing land elevations. (After a map by Dr. Ron Blakey, Northern Arizona University.)

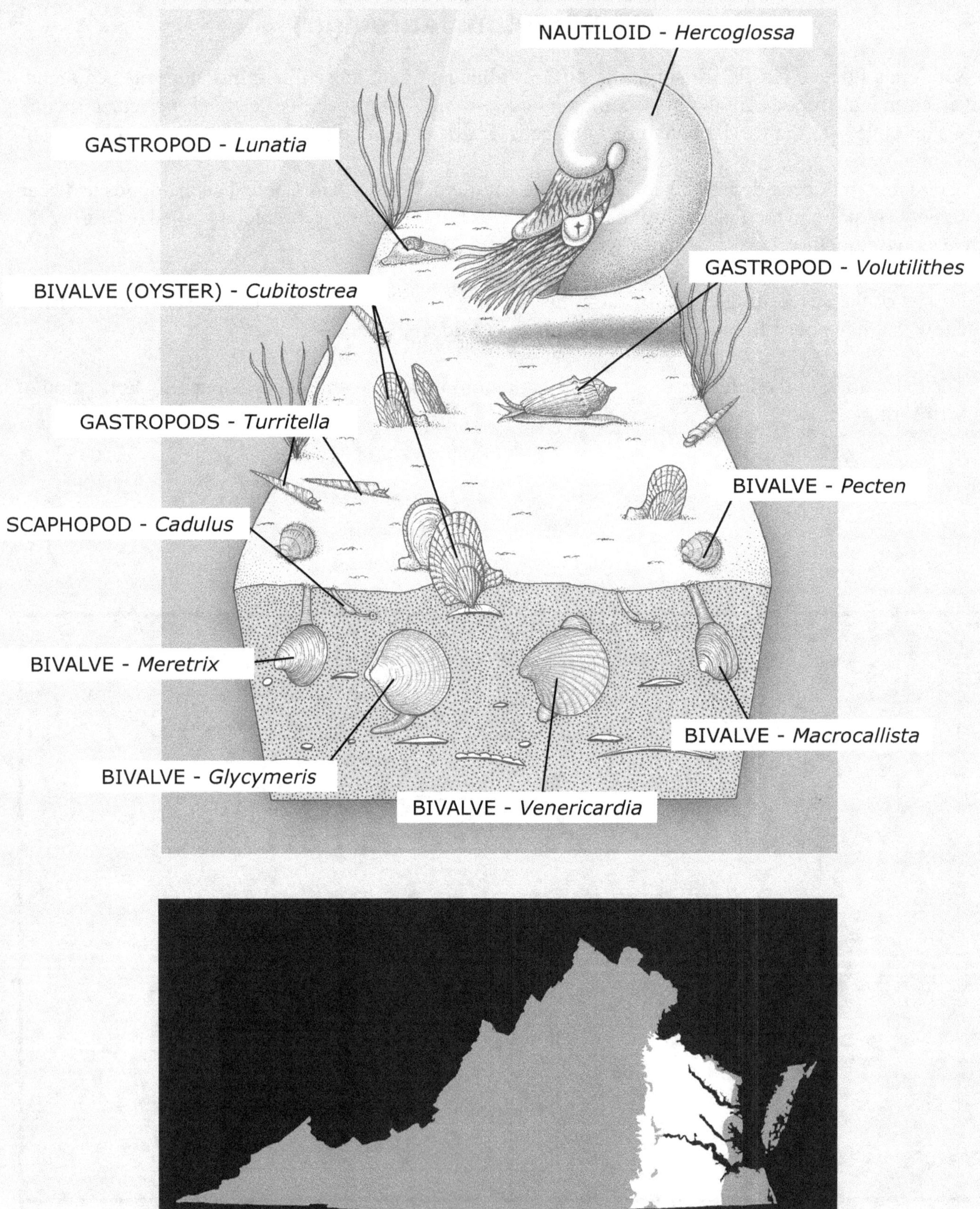

Distribution of surface Tertiary sedimentary rocks and deposits in Virginia (white)

Eocene Fossils of Virginia

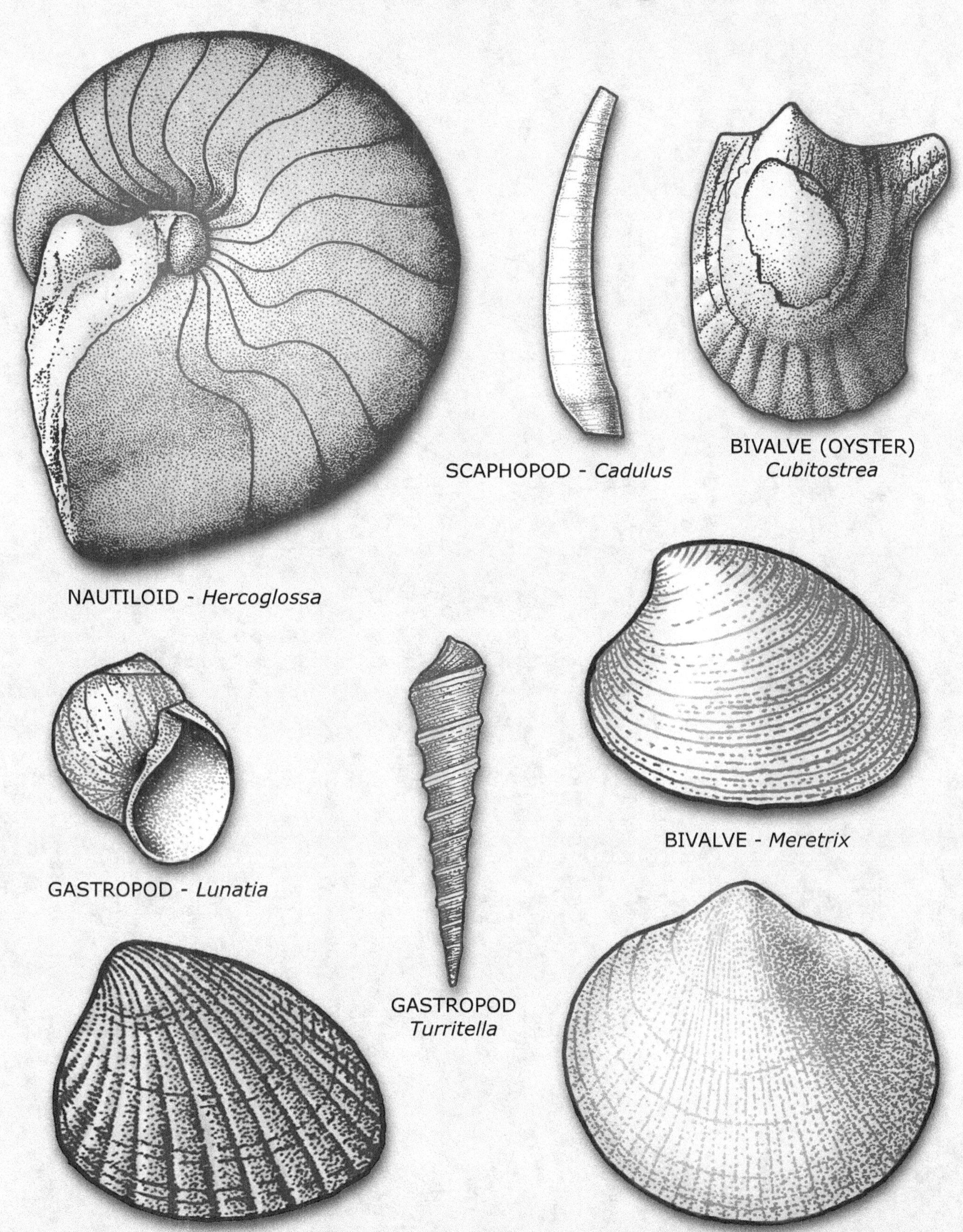

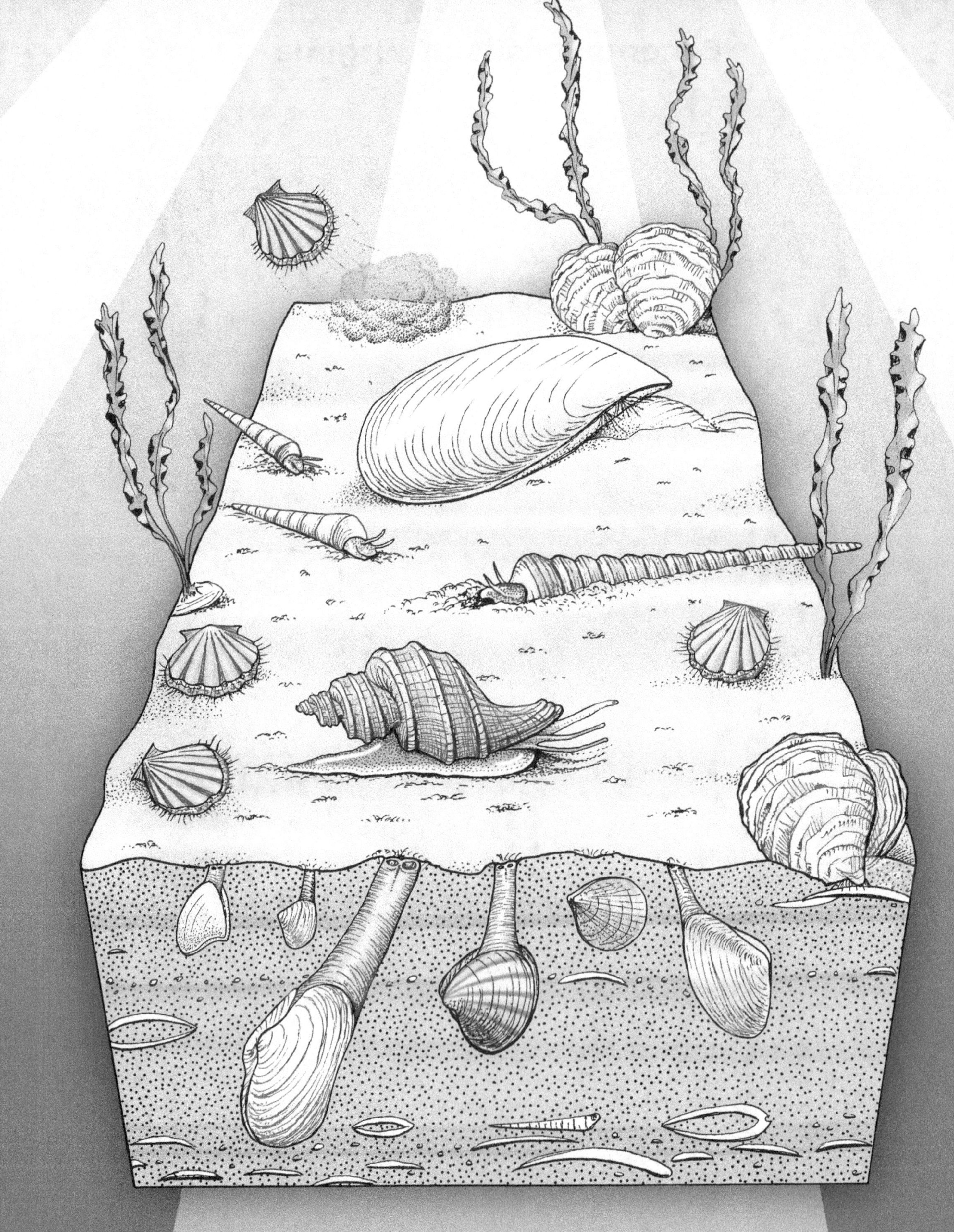

The Miocene Epoch in Virginia
(23-5.3 million years ago)

The Miocene fossil assemblages of Virginia, Maryland, and North Carolina are world-famous for their excellent preservation and staggering variety. In addition to a rich invertebrate fauna, many vertebrate types may be found, such as the bones and teeth of whales, manatees, and crocodiles and the remains of turtles, rays, bony fish, and sharks – including the gigantic *C. megalodon* (*"C"* stands for either *Carcharocles* or *Carcharodon* – the correct name is under dispute), whose teeth may exceed seven inches in slant height! Occasionally, the remains of birds and terrestrial mammals that washed out to sea after death are found, including the teeth and bones of horses and elephant-like gomphotheres.

Sea level was much higher in Virginia during the Tertiary Period than it is today, which is the reason such a rich fossil record is available for study on dry land. The fossils of animals that lived in fairly deep ocean waters may now be found in pits or along beaches because the sea has retreated to the east.

The drawing opposite is based on the fossils found in the Calvert Formation of Virginia.

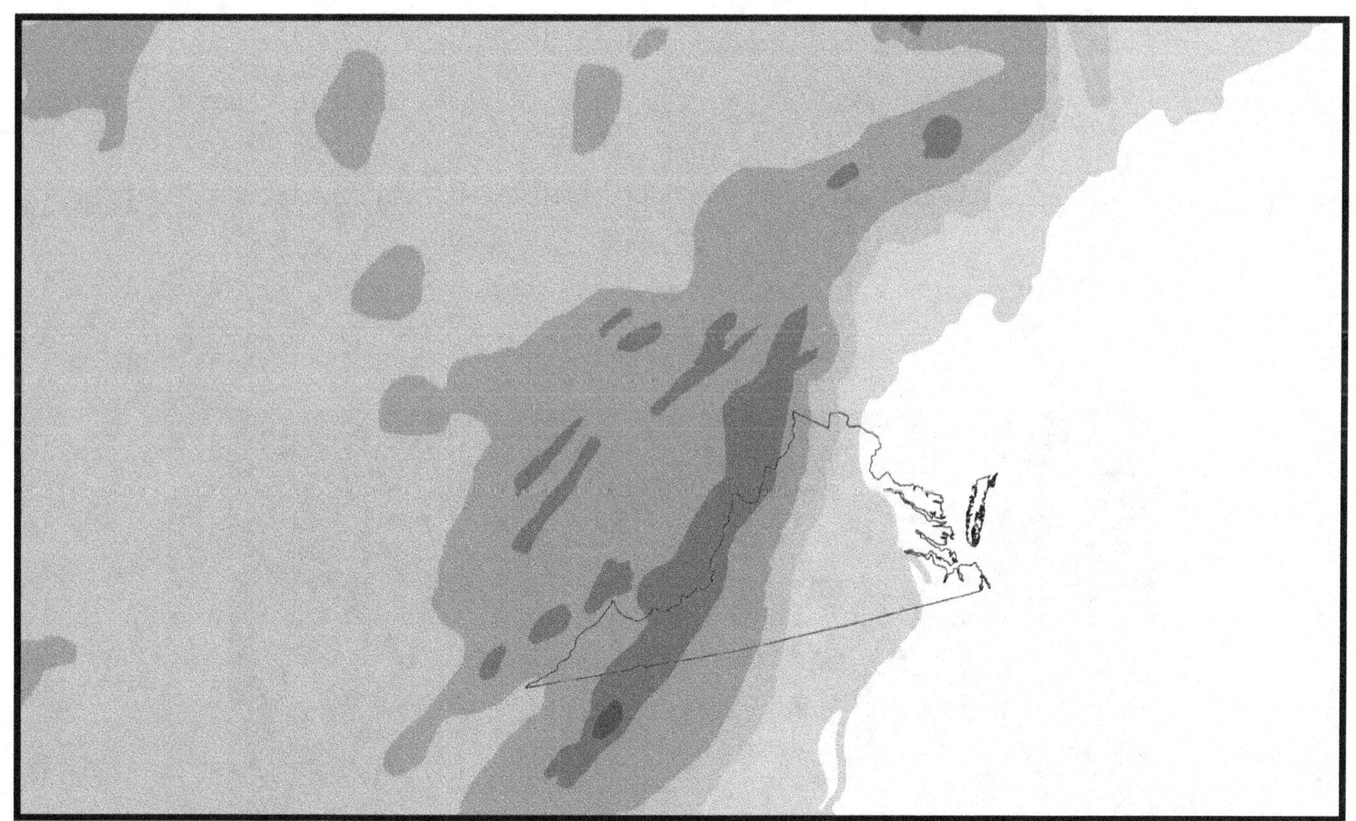

Elevation map of the region of Virginia (outline) during the Miocene Epoch, with white representing water and progressively darker shades of gray representing increasing land elevations.
(After a map by Dr. Ron Blakey, Northern Arizona University.)

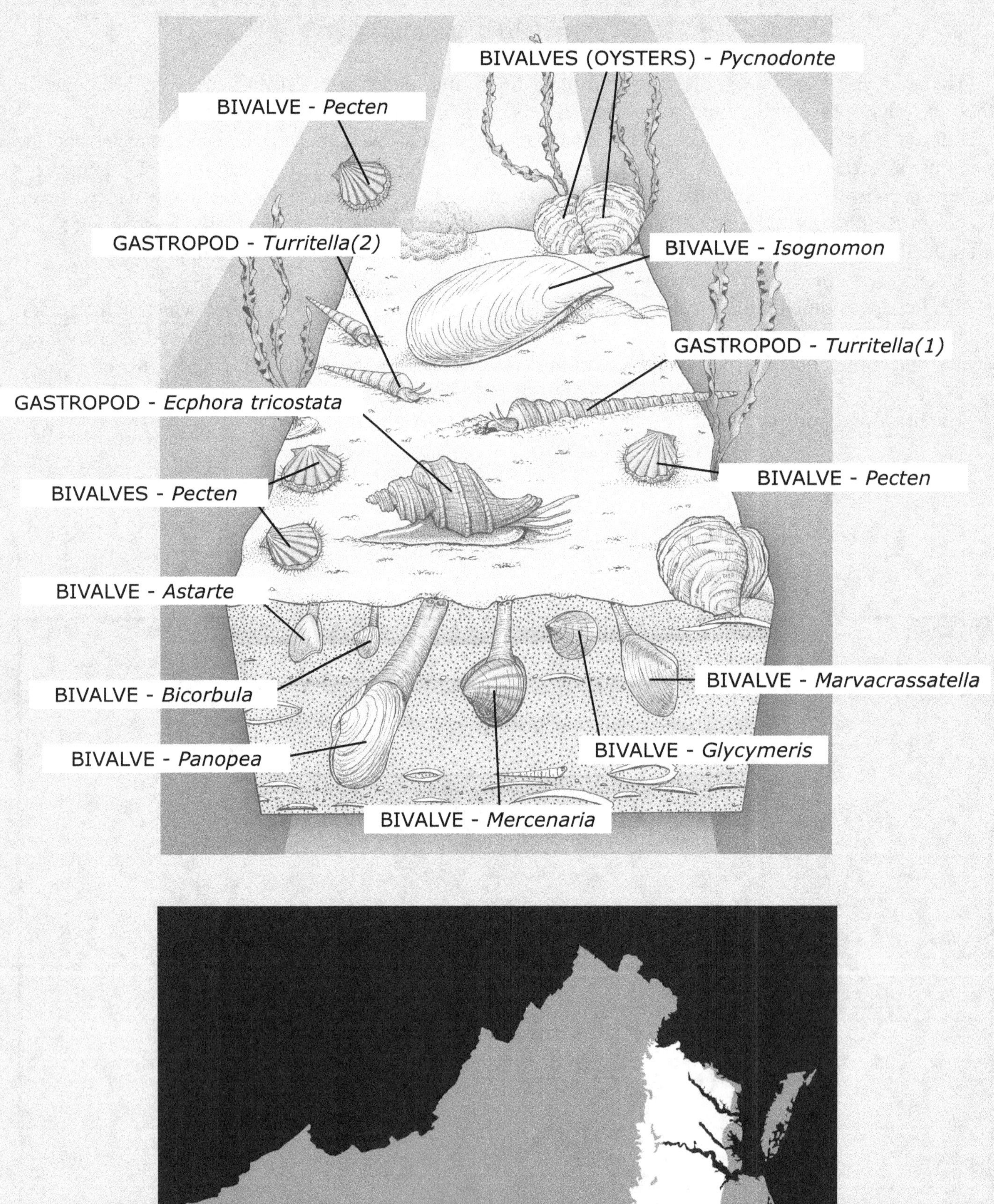

Distribution of surface Tertiary sedimentary rocks and deposits in Virginia (white)

Miocene Fossils of Virginia

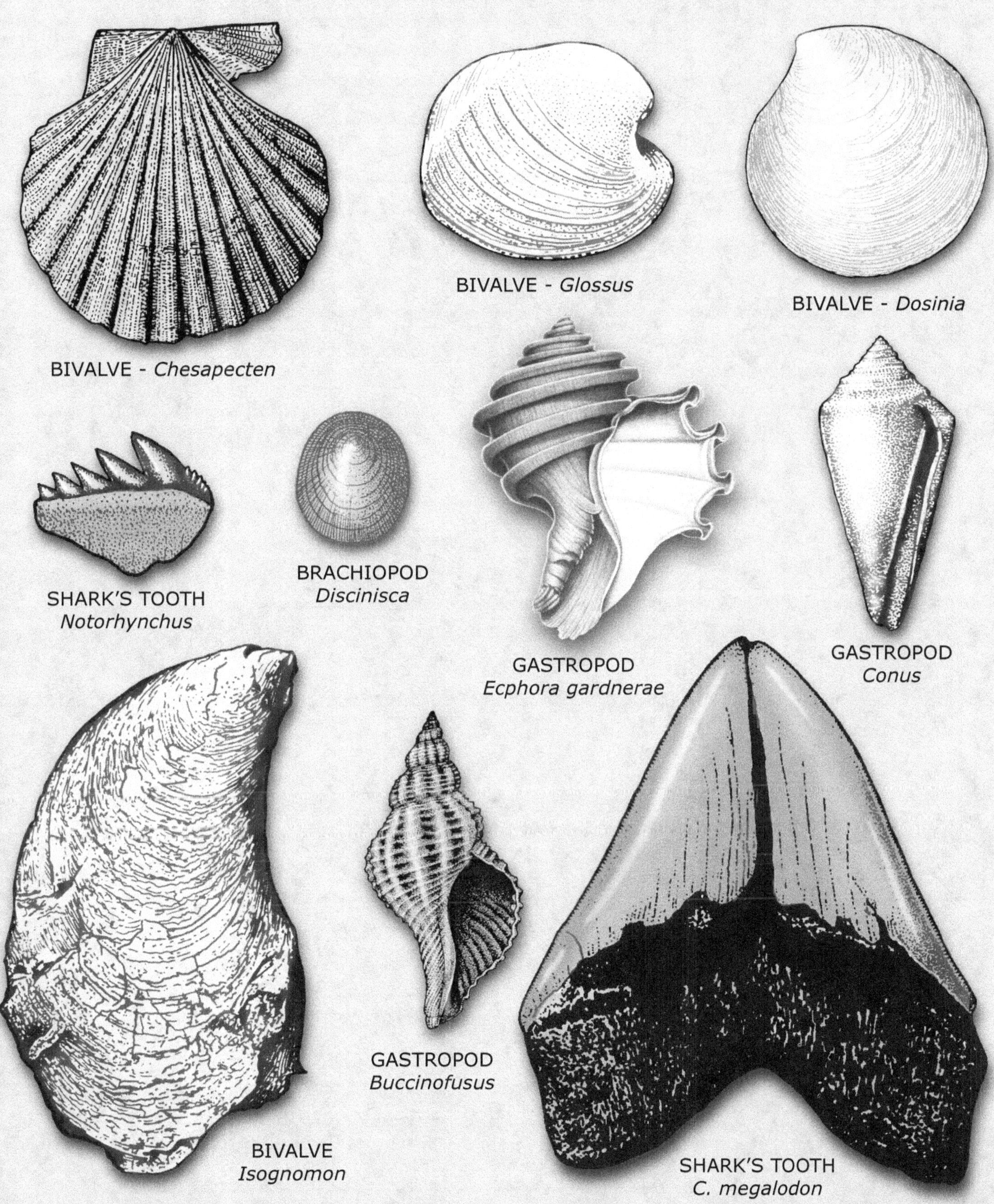

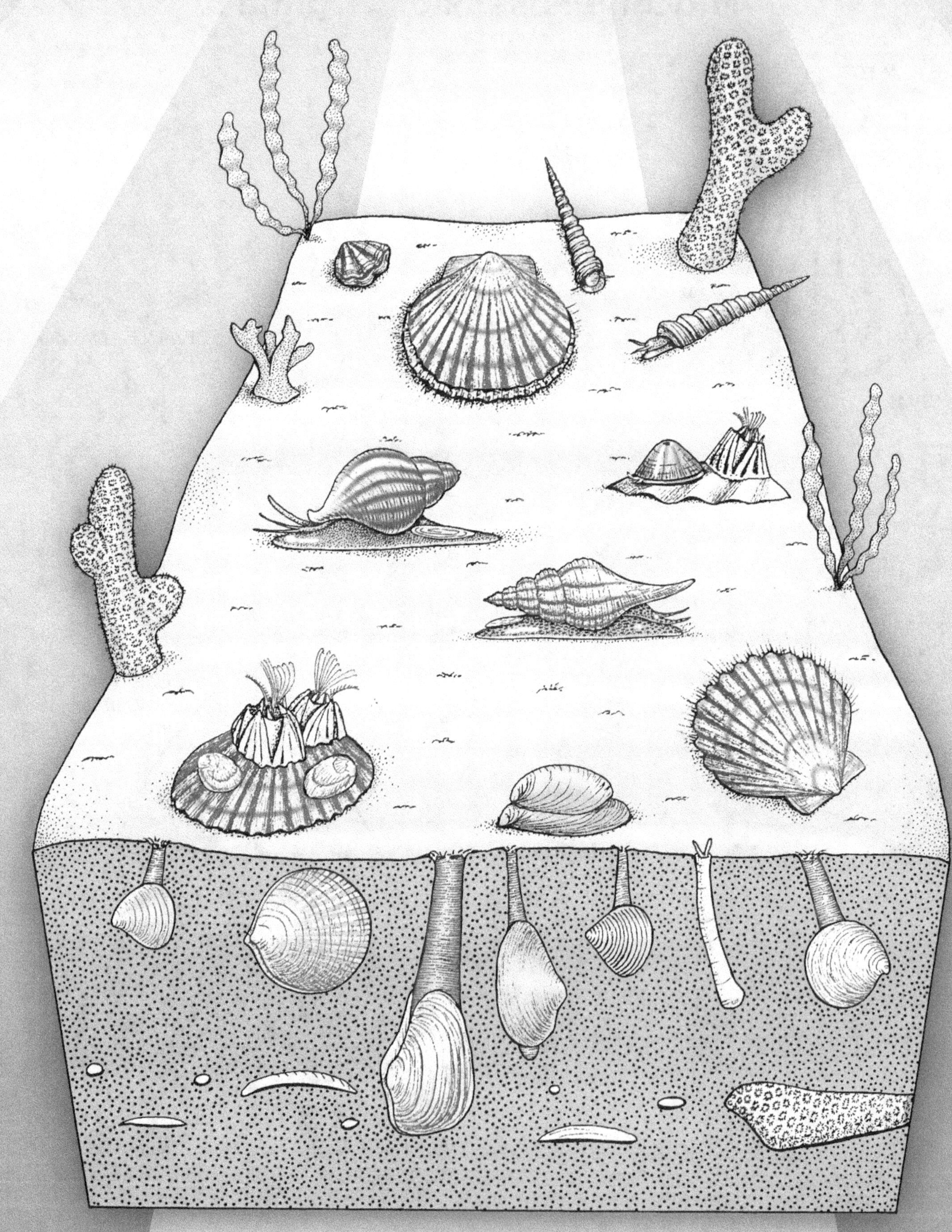

The Pliocene Epoch in Virginia
(5.3-2.6 million years ago)

The abundance and excellent preservation of Miocene fossils in Virginia continues in the Pliocene deposits. Bivalves, gastropods, corals, whale and porpoise bones and teeth, sharks' teeth, and many other remains may be found. The presence of numerous large coral fossils is evidence that Virginia was much warmer during the early Pliocene Epoch than it is today.

One of the more impressive invertebrate fossils from the Virginia Pliocene is the bivalve (extinct scallop) *Chesapecten jeffersonius*, named in honor of Thomas Jefferson and designated as the official state fossil of Virginia. These shells may be as much as nine inches across.

The diorama on the facing page shows the marine organisms that are preserved in the Yorktown Formation in Virginia.

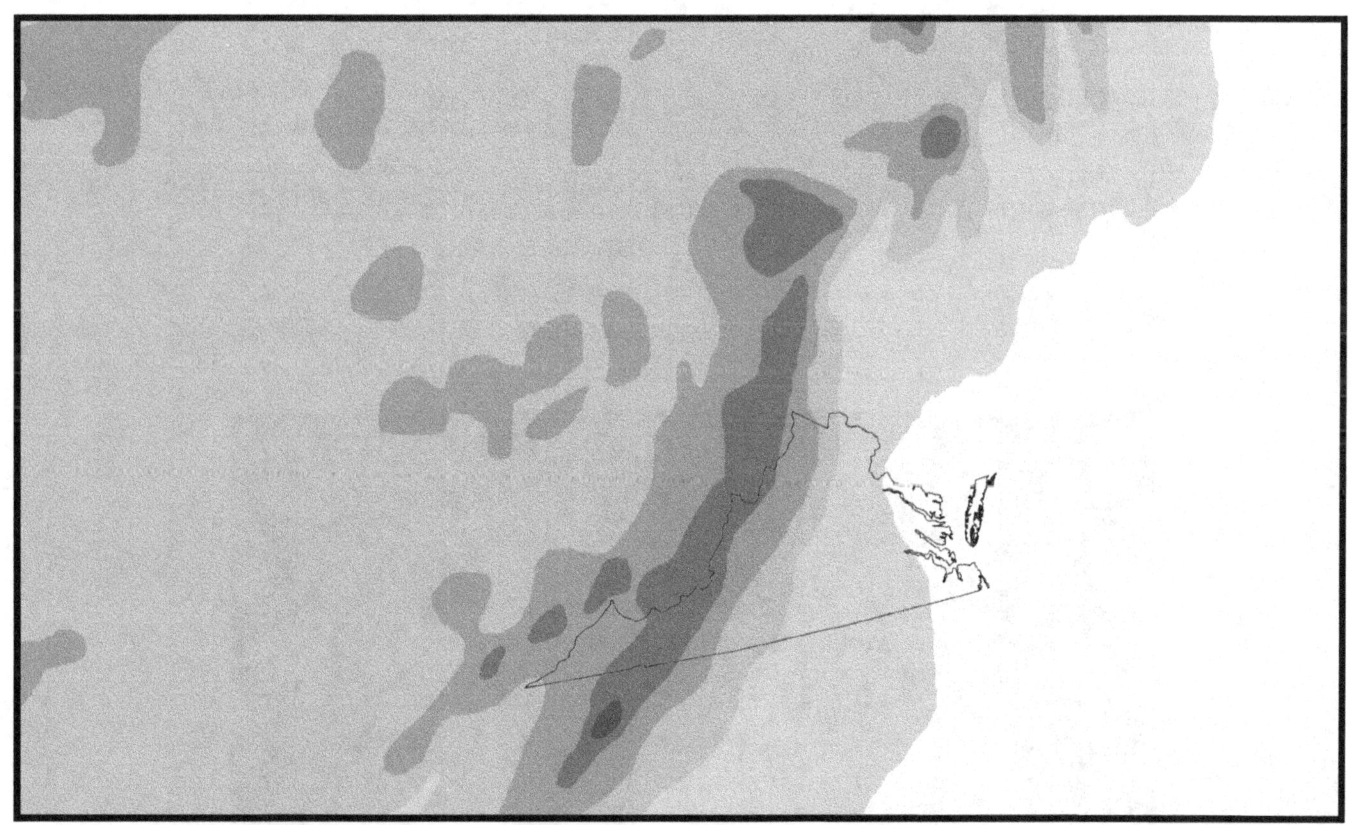

Elevation map of the region of Virginia (outline) during the Pliocene Epoch, with white representing water and progressively darker shades of gray representing increasing land elevations.
(After a map by Dr. Ron Blakey, Northern Arizona University.)

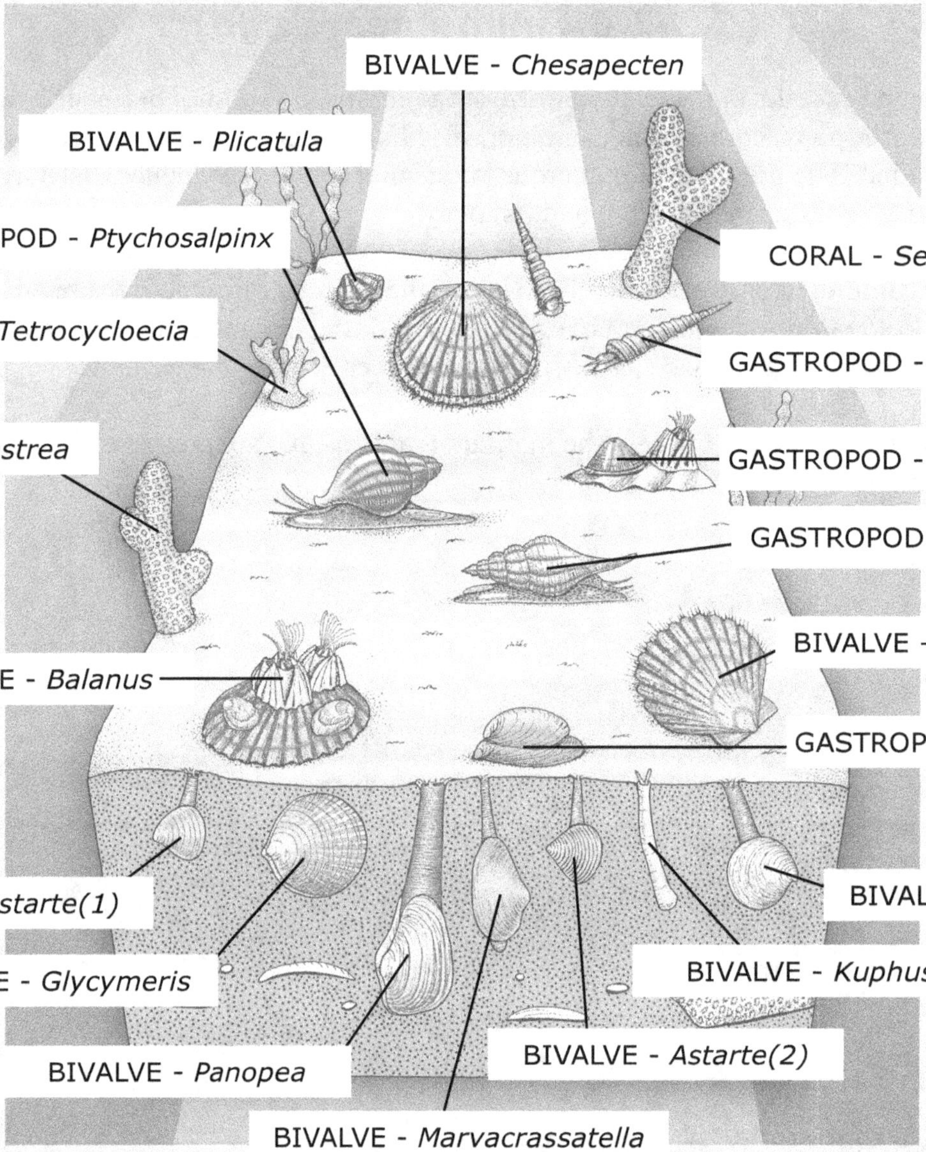

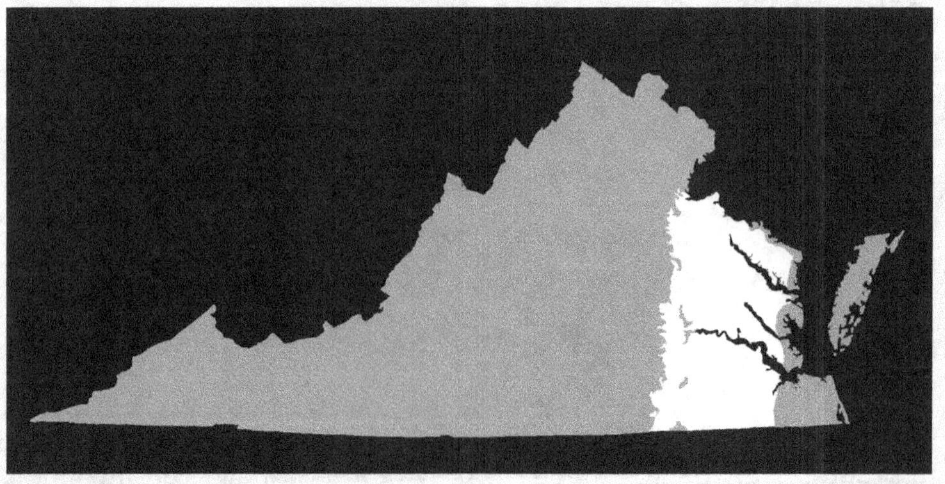

Distribution of surface Tertiary sedimentary rocks and deposits in Virginia (white)

Pliocene Fossils of Virginia

The Pleistocene Epoch in Virginia
(2.6 million to 11,700 years ago)

In the late Pliocene and Pleistocene Epochs, the climate in Virginia became much colder. In fact, the Pleistocene is often called "The Ice Age". The northern polar ice cap expanded greatly at this time, reaching as far south as Pennsylvania and Kentucky (but not Virginia itself) and bringing frigid weather with it. The kinds of animals and plants that lived in the region changed and sea level dropped considerably as much water was held in the form of ice. The glaciers advanced and retreated several times during the Pleistocene, with some intervals being even warmer than today.

Pleistocene marine fossils are scarce in Virginia as most marine deposits accumulated in what is now deep water off the coast. Some collecting sites occur in the southeastern corner of the state and contain bivalves, gastropods, sand dollars, shark teeth, whale bones, and other remains.

Terrestrial fossils of Pleistocene age are not common in Virginia, but a great variety of types have been discovered. Pleistocene deposits may be found along rivers and lakes, and in some isolated sites like caves and low-lying valleys and canyons. Plant specimens that have been found include pieces of petrified wood and fossilized pine cones and pecans. Mammal fossils from the state from this time include the familiar Ice Age mammoths, mastodons, and giant ground sloths. The remains of many other animals have been found as well: bears, tapirs, beavers, crocodiles, walruses, horses, camels, deer, moose, musk ox, and bison. In addition, artifacts created by early Native Americans are frequently found that date from the latest Pleistocene.

The scene opposite is based on the discoveries of late Pleistocene (roughly 14,500 years old) mammal fossils near Saltville, in the southwestern part of the state.

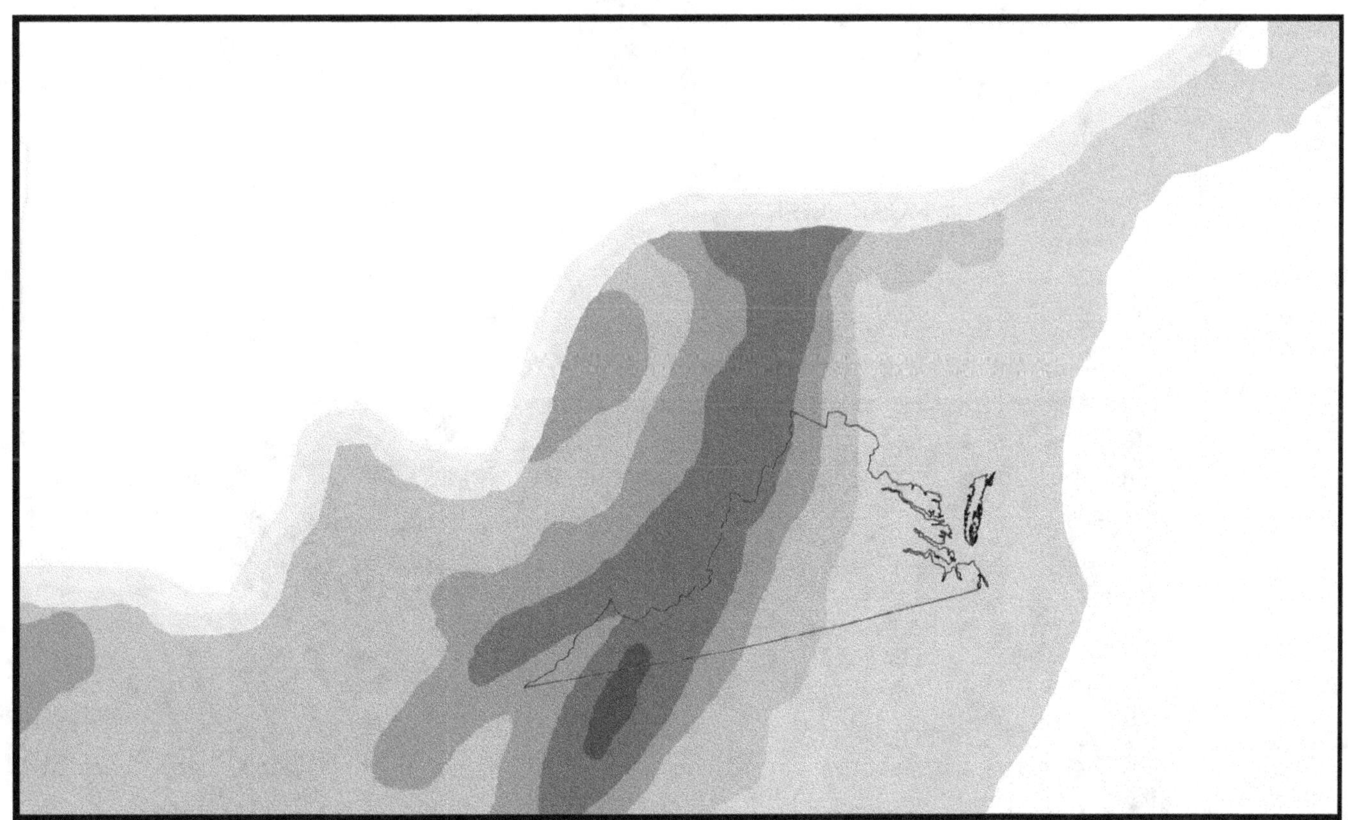

Elevation map of the region of Virginia (outline) during the Pleistocene Epoch: white at top represents glacial ice; white at right represents water; progressively darker shades of gray represent increasing land elevations. (After a map by Dr. Steven Dutch, Univ. of Wisconsin - Green Bay.)

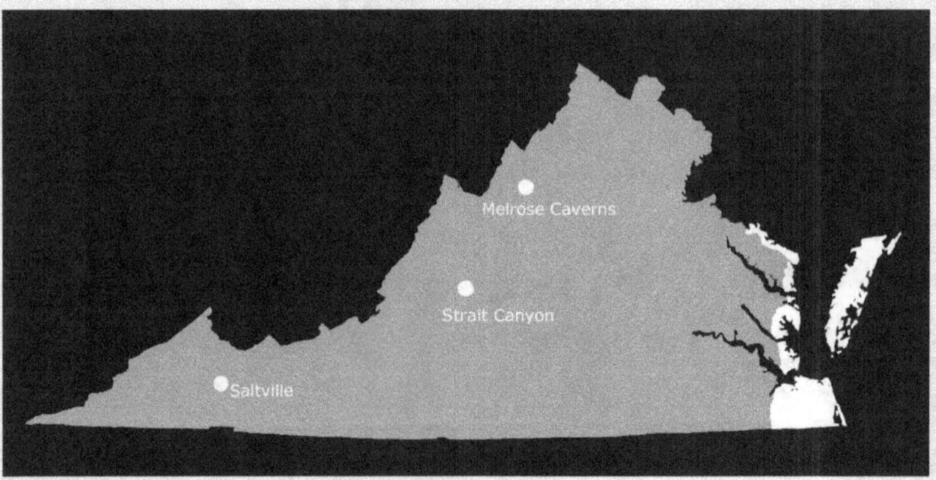

Distribution of surface Pleistocene sedimentary rocks and deposits and some isolated fossil sites in Virginia (white)

Pleistocene Fossils of Virginia

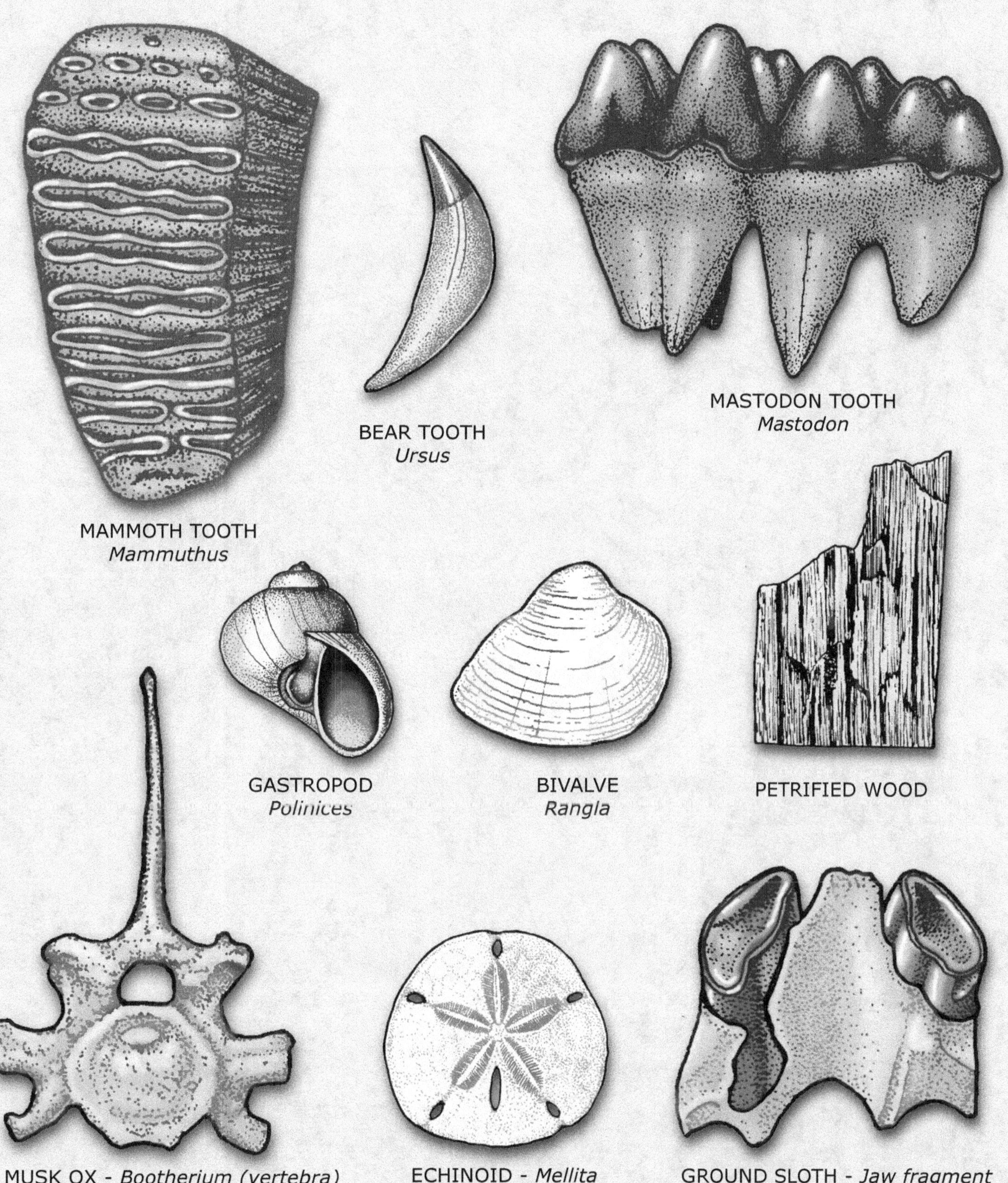

The Holocene Epoch in Virginia
(11,700 years ago to the Present)

It has been said that before human beings came to eastern North America the forest was so thick and extensive that a squirrel could, if so inclined, travel all the way from Maine to Florida by jumping from tree to tree, without ever having to touch ground. That forest was chopped down long ago by man – there is almost no virgin forest left in the state today. Many other changes have also been brought by human settlers, from the damming of rivers to the introduction of new, high-impact species from other parts of the world, such as dandelions, Japanese beetles, starlings, pigeons, and Norway rats.

Though the impact of human activities on the natural landscape of Virginia has been profound, it should be remembered that animals and plants still dominate, both in numbers and in total biomass. There are more critters in any shovel full of top soil than there are people in the entire state. (For that matter, there are more critters living inside each human being than there are people living in Virginia.) The return of many wild animals to urban and suburban environments (e.g. white-tailed deer, red foxes, raccoons, opossums) demonstrates their adaptability and resourcefulness. They may well continue to prowl the land long after human beings are gone.

The Earth is a dynamic system – change is constant and unrelenting. A quick browse through the maps and dioramas in this book may give some idea of just how much the land we call Virginia has changed over time, both as a physical environment and in the kinds of plants and animals that have lived here. The current situation and cast of living characters are temporary - just one chapter in a continuing story.

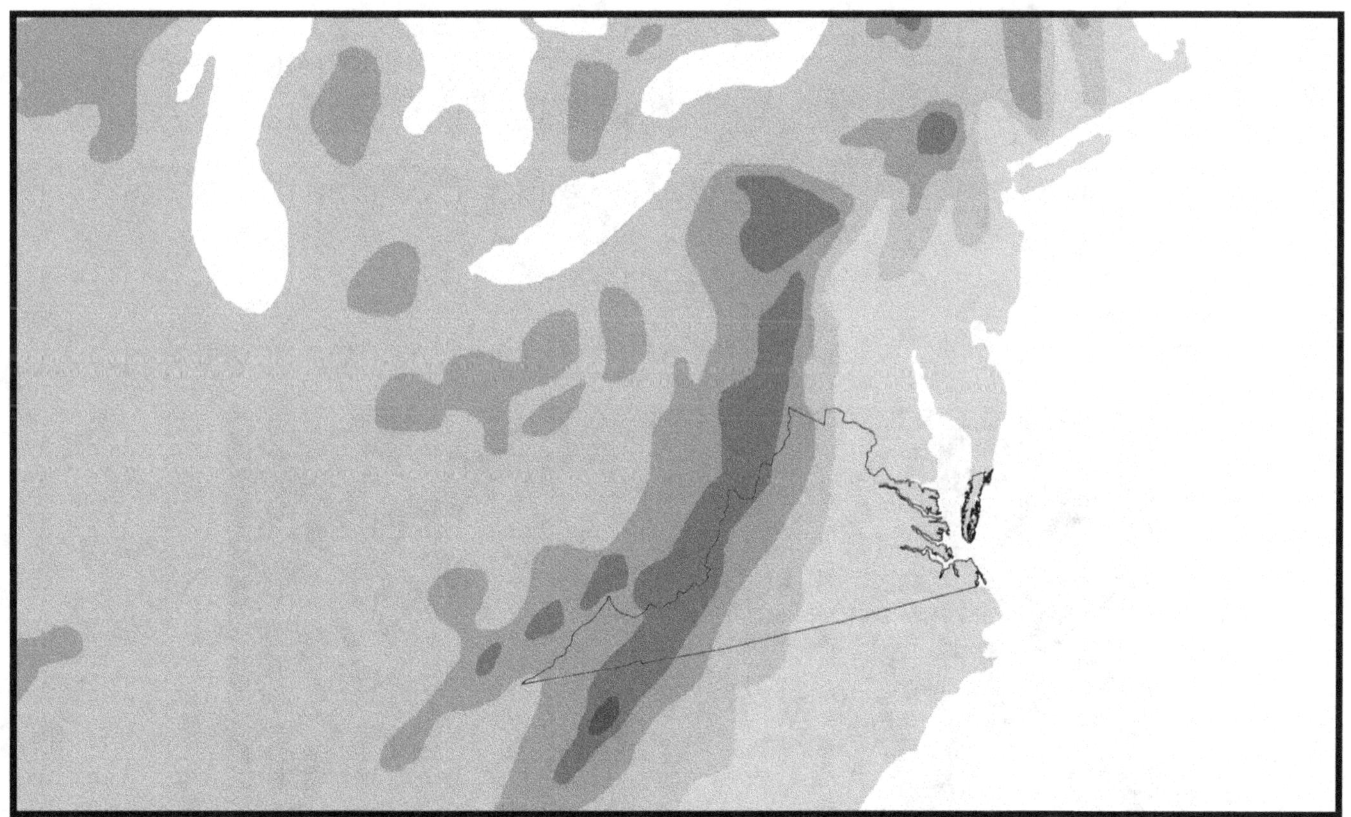

Simplified elevation map of present-day Virginia, with white representing water and progressively darker shades of gray representing increasing land elevations.

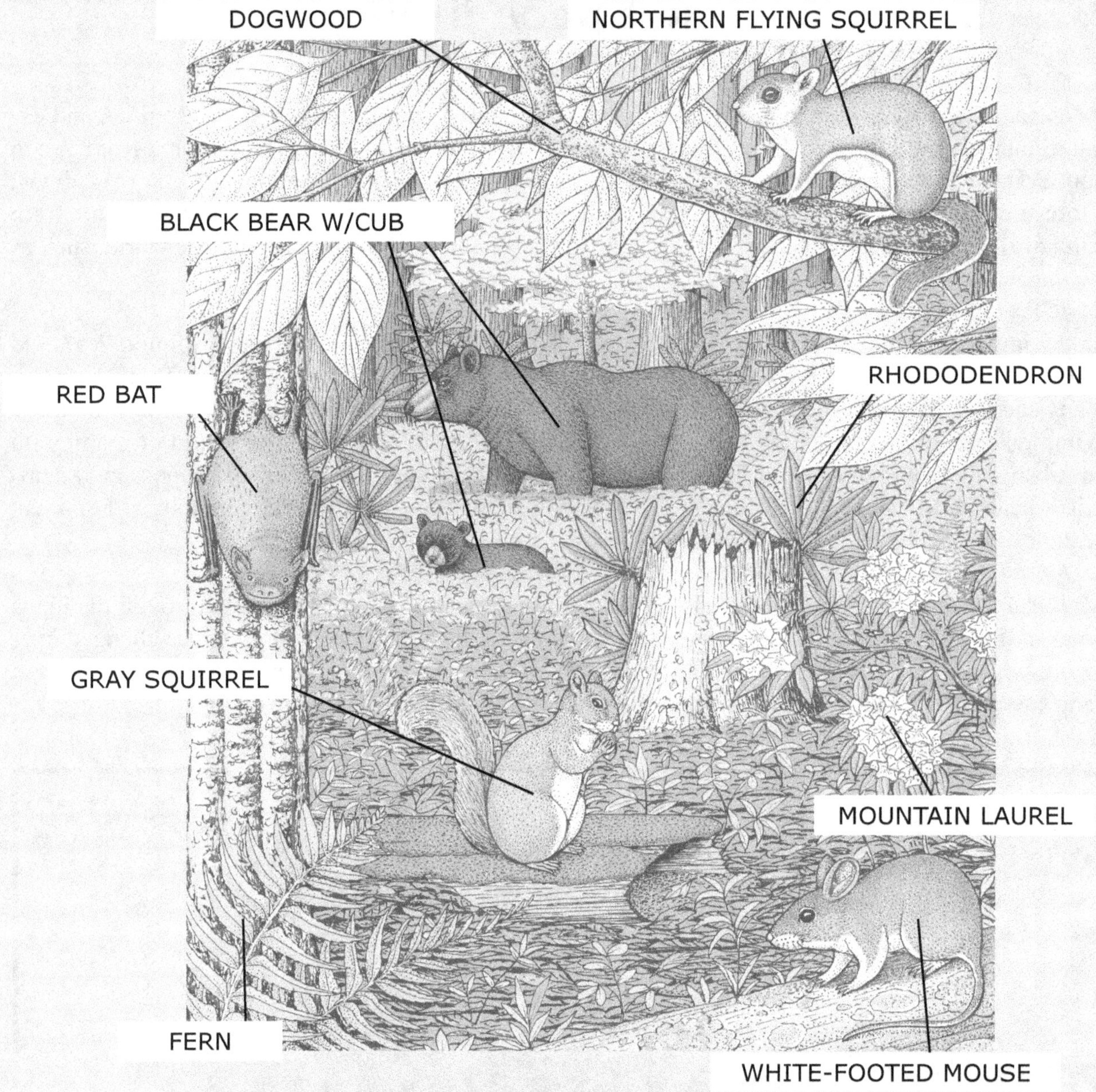

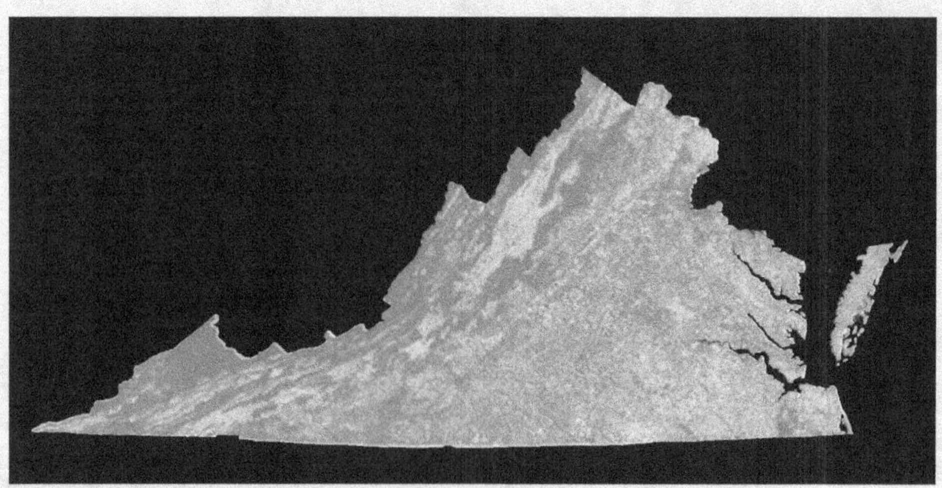

Distribution of forests (dark) in present-day Virginia (Source: Va. Dept. of Forestry)

Living Plants and Fungi of Virginia

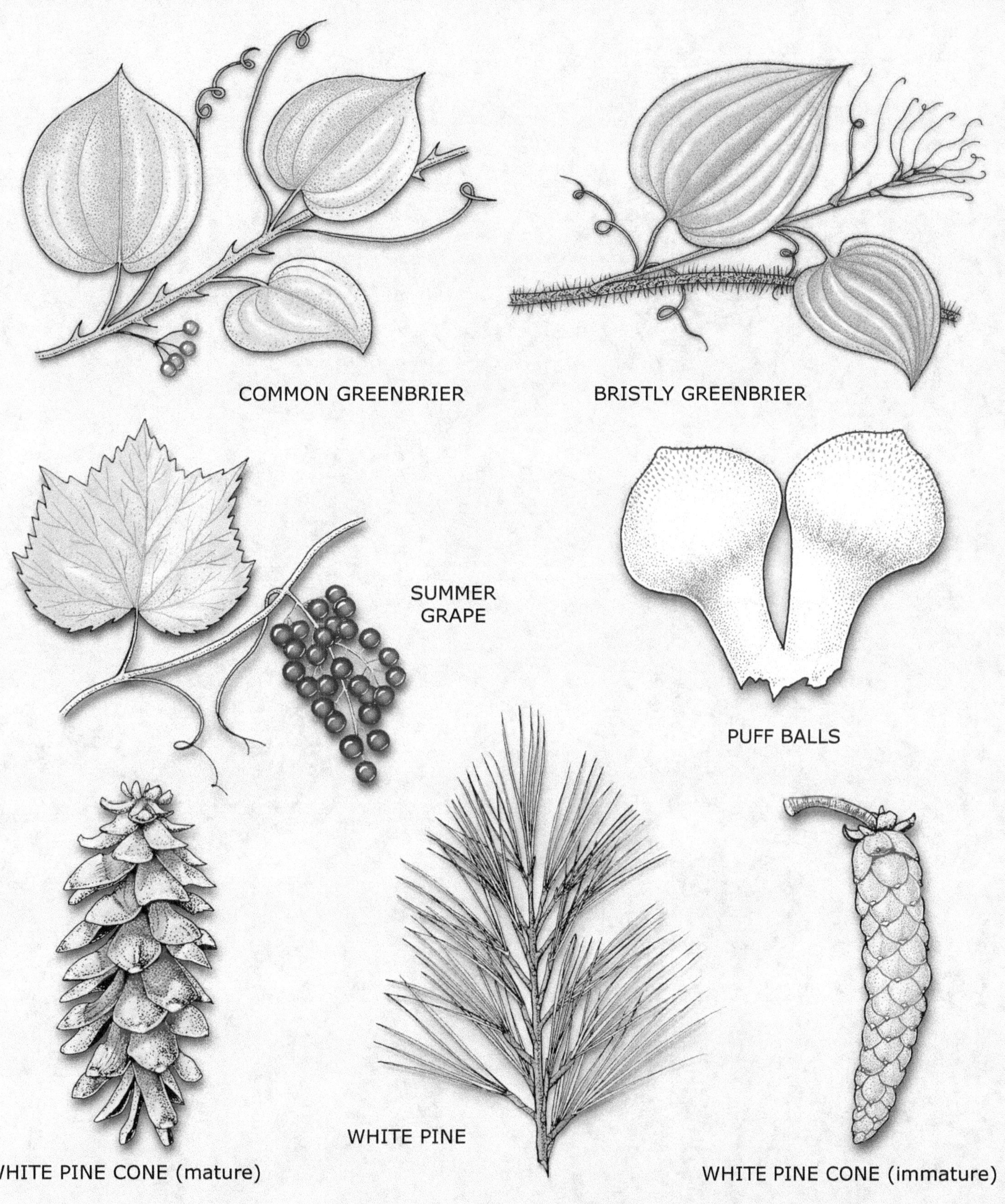

Living Plants and Invertebrates of Virginia

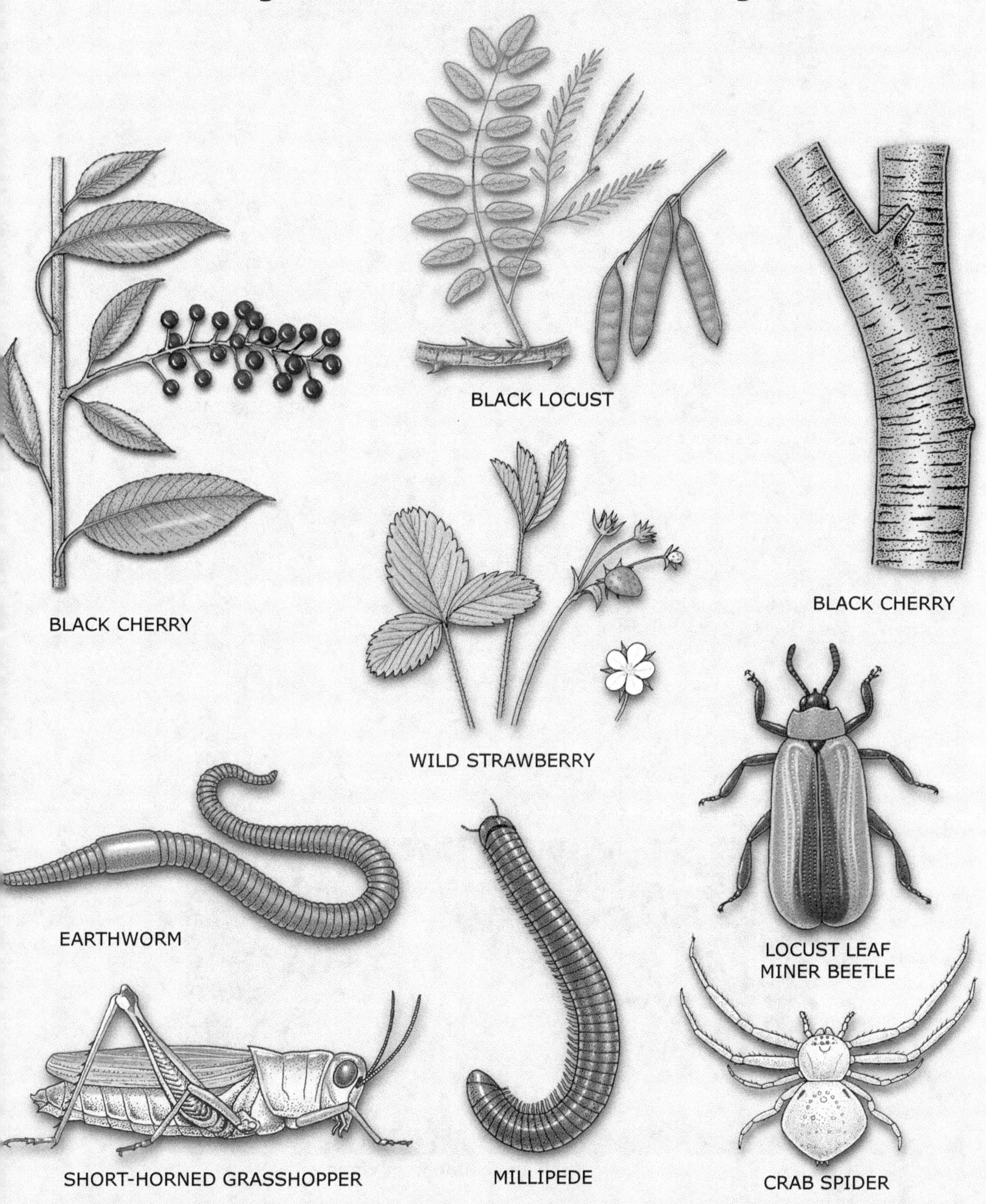

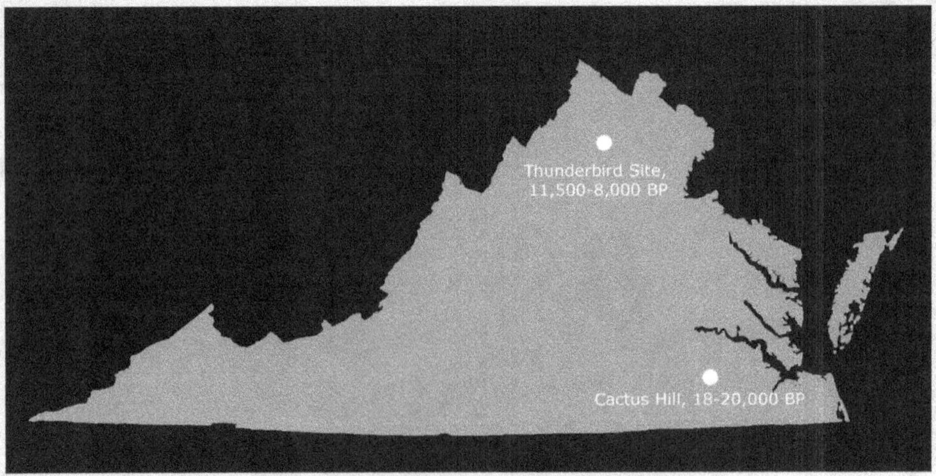

Early sites of Indian habitation in Virginia

Living Invertebrates of Virginia

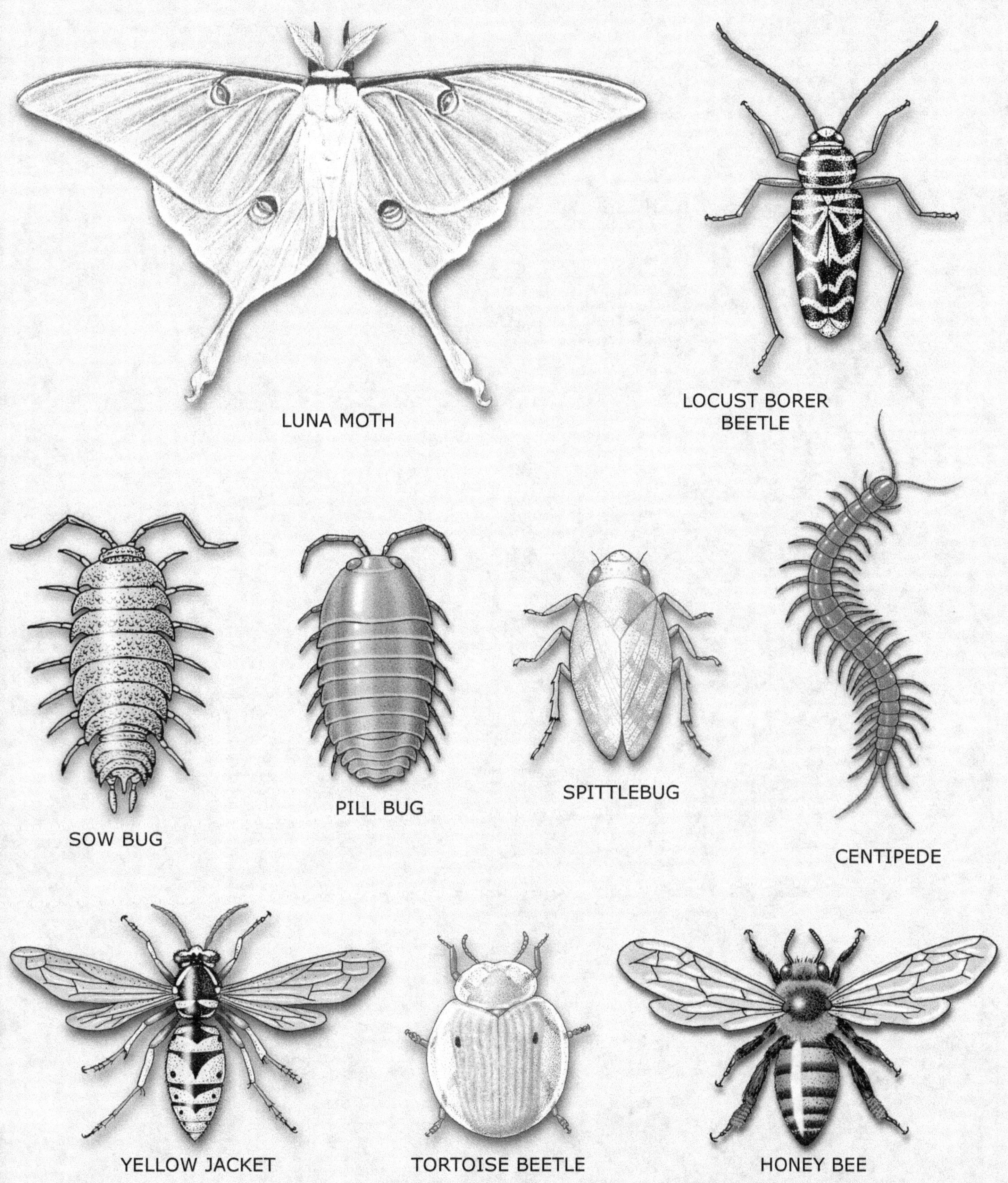

Living Vertebrates of Virginia

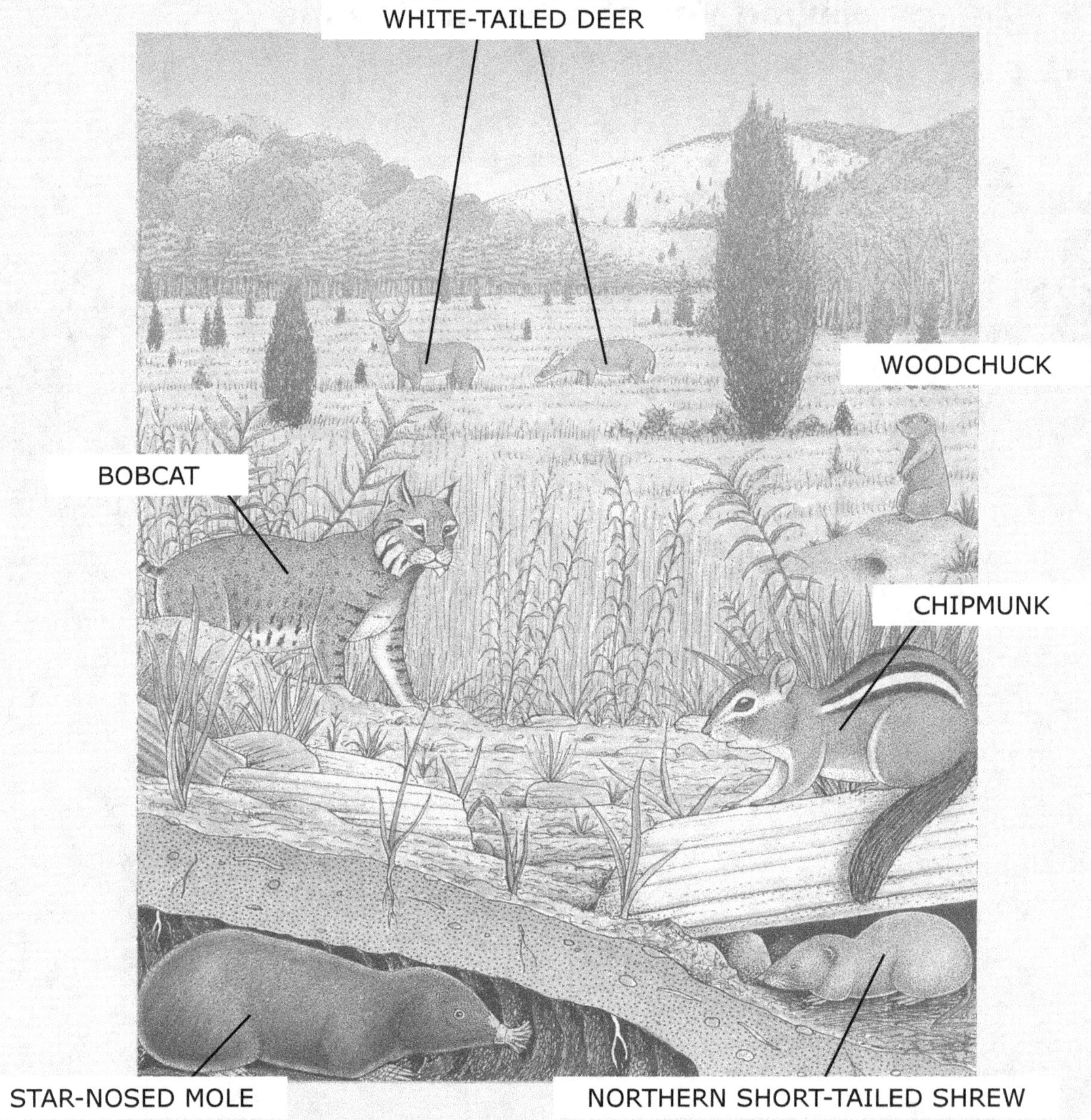

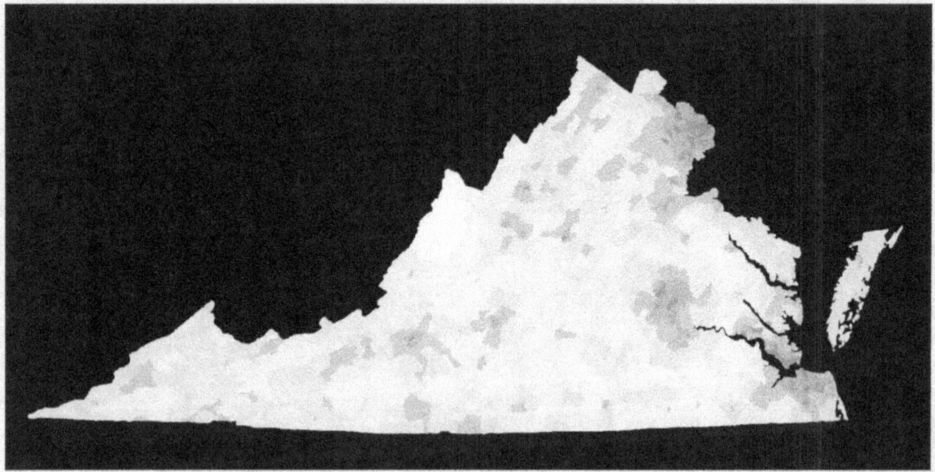

Human Population distribution in present-day Virginia (darker gray = more people)

Living Vertebrates of Virginia

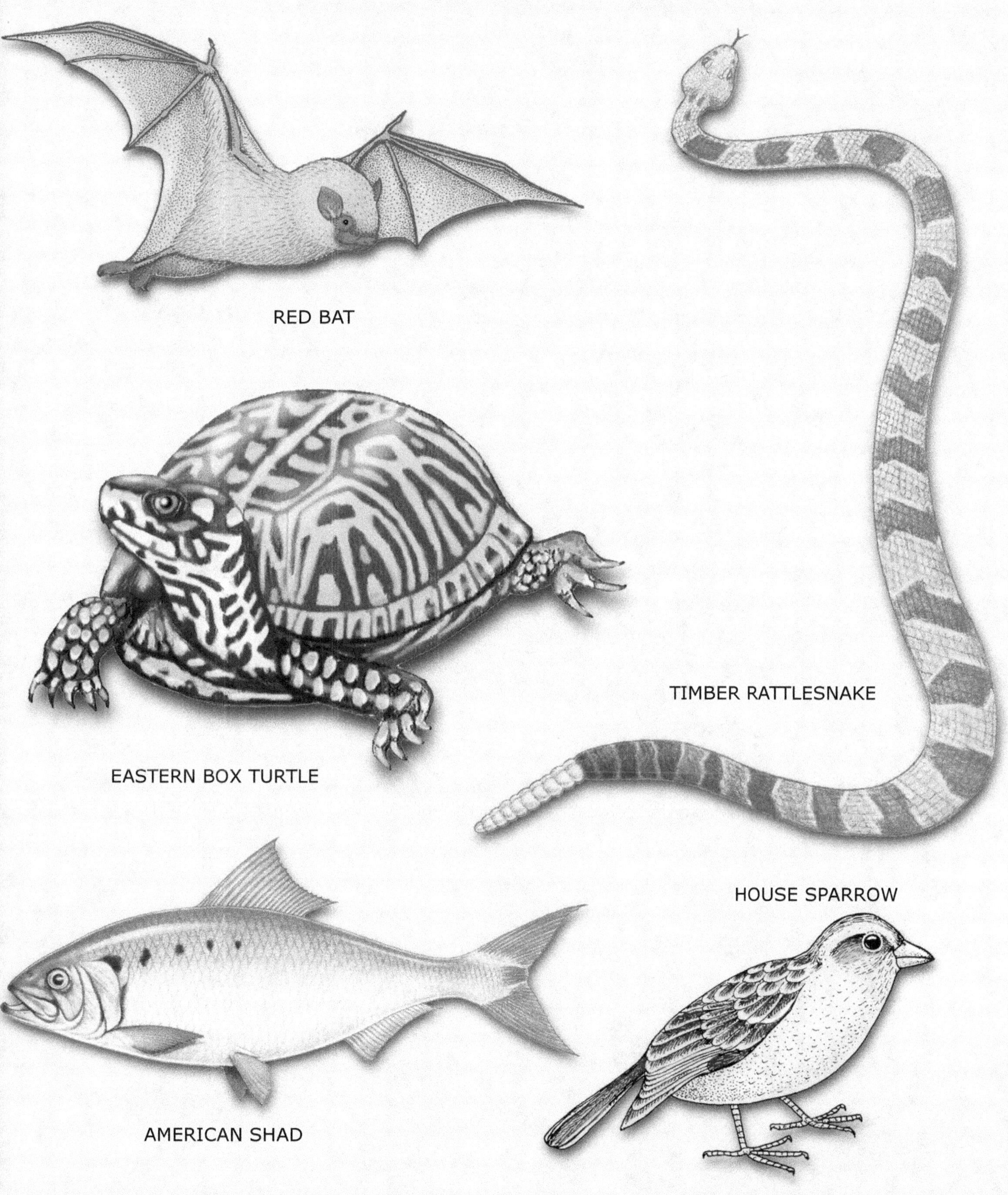

More by Jasper Burns

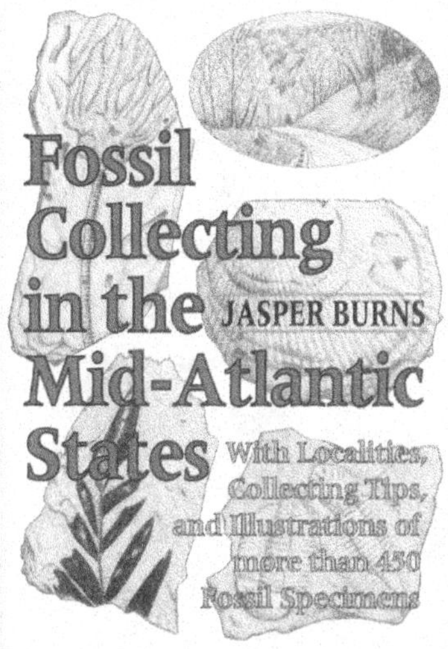

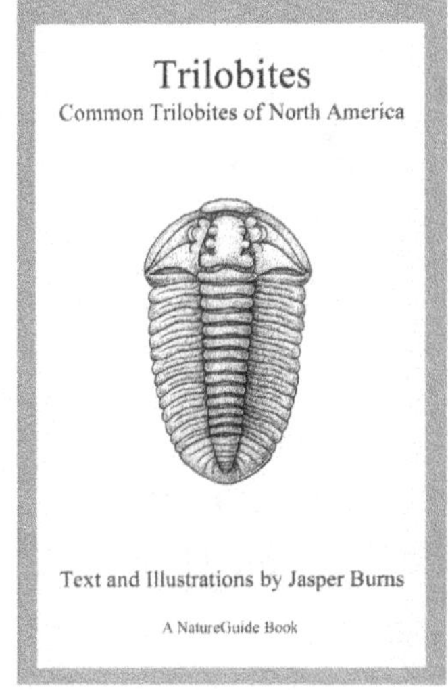

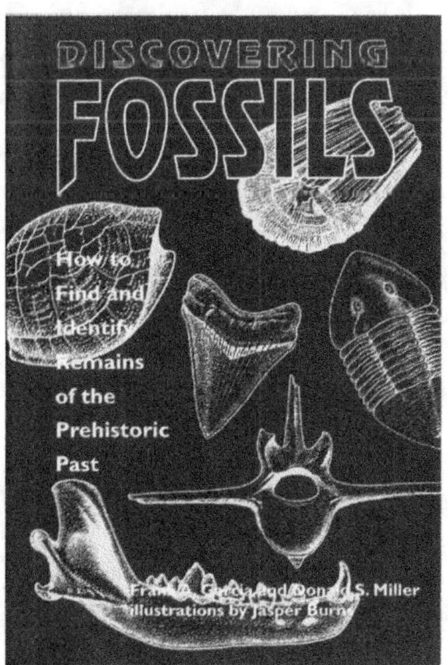